Feb 2010
$40⁰⁰
NFIC
363.7
CHA

Industrial Chemicals & Health

Health and the Environment
Books in This Series

Air Pollution & Health
Global Warming & Health
Industrial Chemicals & Health
Population Growth & Health
Water Pollution & Health
You & the Environment

Industrial Chemicals & Health

Zachary Chastain

AlphaHouse Publishing
New York

Health and the Environment
Industrial Chemicals & Health

Copyright © 2009 by AlphaHouse Publishing, a division of PEMG Publishing Group. All rights reserved. No part of this publication may be reproduced or transmitted in any form or by any means, electronic or mechanical, including photocopying, recording, taping, or any information storage and retrieval system, without permission from the publisher.

AlphaHouse Publishing
A Division of PEMG Publishing Group, Inc.
201 Harding Avenue
Vestal, New York 13850
www.alphahousepublishing.com

First Printing
9 8 7 6 5 4 3 2 1
ISBN: 978-1-934970-37-9
ISBN (set): 978-1-934970-34-8

Library of Congress Control Number: 2008930664
Author: Chastain, Zac

Cover design by Wendy Arakawa.
Interior design by MK Bassett-Harvey.

Printed in India by International Print-O-Pac Limited
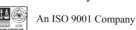 An ISO 9001 Company

Contents

Introduction   6
1. What Are Industrial Chemicals   9
2. Who Is at Risk?   27
3. Pesticides   39
4. Dioxins, PCBs, and Phthalates   53
5. Common Toxic Chemicals and Their Health Effects   67
6. What Is the World Doing?   81
For More Information   90
Glossary   92
Bibliography   109
Index   110
Picture Credits   111
About the Author and the Consultant   112

Introduction

"The word *environment* does not mean something that surrounds us but an organism of all life within which we are fastened."
—Mose Richards, in
The Cousteau Almanac (1980)

Discussing the environment, we tend to speak as though it were a separate entity. "Protect the environment!" we demand, overlooking that *we* are an inseparable part of the environment. The air we breathe, the water we drink, the trash we discard, the sunlight to which we're exposed—all are different aspects of our environment. This series educates readers about humans' place in the environment and describes how intertwined our lives are with the natural world. Readers will also learn how certain human activities are degrading our environmental life-support systems, disrupting natural ecosystems and endangering human health in the process.

Among those at greatest risk for serious illness due to environmental pollution are children, the elderly, and those living in poverty, with children bearing the greatest share of the burden. According to the United Nations Environmental Program (UNEP), the quality of a child's environment is one of the key factors as to whether he or she will survive the first few years of life, particularly in developing countries. The World Health Organization (WHO) cites statistics showing that each year more than three million children under the age of five die due to environment-related diseases. Those claiming the highest toll include:

- Acute respiratory infections, 60% of which are related to environmental conditions.
- Diarrheal diseases, 80%-90% of which are a result of contaminated drinking water and poor sanitation.
- Malaria, the vast majority of cases resulting from lack of adequate mosquito control.

Such problems can only be solved through better environmental management programs. According to the WHO, preventing environmental disease could save the lives of as many as four million children each year. This series provides information on the health risks posed by environmental pollutants, describes ongoing prevention and management programs around the world, and offers useful advice to readers on how they can reduce their own risk of environmental disease.

Not only does this series describe how the quality of the environment affects us—it also explains how human activities affect the environment. Pollutants from automobile and factory emissions foul the air while wastewater from industrial discharges and inadequately treated sewage contaminate waterways. Improper disposal of hazardous materials and municipal solid waste can contaminate soils and groundwater, while careless use of chemical pesticides presents hazards to both humans and wildlife. Pollutants encountered in tobacco smoke, food, and water have been blamed for an increase in cancer rates and intestinal diseases. Humans create chemicals and wastes that foul the environment and these pollutants, in turn, can make humans sick. In addition, human population growth puts a steadily increasing strain on ecosystems, making environmental cleanup even more of a challenge.

Perhaps the greatest environmental threat of all is global warming. The predicted increase in violent weather, droughts, rising sea levels, spread of insect carriers of disease, crop failures, and resulting civil unrest will have a profoundly negative effect on people and the environment everywhere in the world.

We all need to learn how to make changes now to prevent irreparable damage to our atmosphere and our planet before it's too late. This series explores how the creation of new government policies and environmental laws is helping to bring about change and discusses how individuals can have a positive impact in their own homes and communities.

— Anne Nadakavukaren

Here's what you need to know

- Chemicals occur naturally in the environment. They are not always harmful, and some (such as water) are vital to our survival.
- Chemicals have been produced in small quantities for thousands of years, but it wasn't until the nineteenth century, when the Industrial Revolution began, that they were produced in large quantities.
- The chemical industry continued to grow alongside other industries—automotive, textile, plastics—because it provided them with chemicals necessary to make their products.
- The chemical industry is one of the largest in the world. It employs millions and is an essential part of our world economy.
- Chemicals become dangerous when they become hazardous waste. Hazardous waste includes chemicals that are corrosive, ignitable, reactive, or toxic.
- Toxic chemicals are substances that can damage organs or disrupt biochemical processes away from the site on the body where exposure occurred.
- There are three types of exposure to toxic chemicals: inhalation, ingestion, and dermal.
- The first priority with toxic chemicals is reduction, then reuse or recycling back into the industry from which they came. If that isn't possible, they should be disposed of as safely as possible.
- The Agency for Toxic Substances and Disease Registry (ATSDR), along with the EPA, is responsible for keeping records of all toxic substances. Unfortunately, they still know very little about the harmful effects of hundreds of substances.

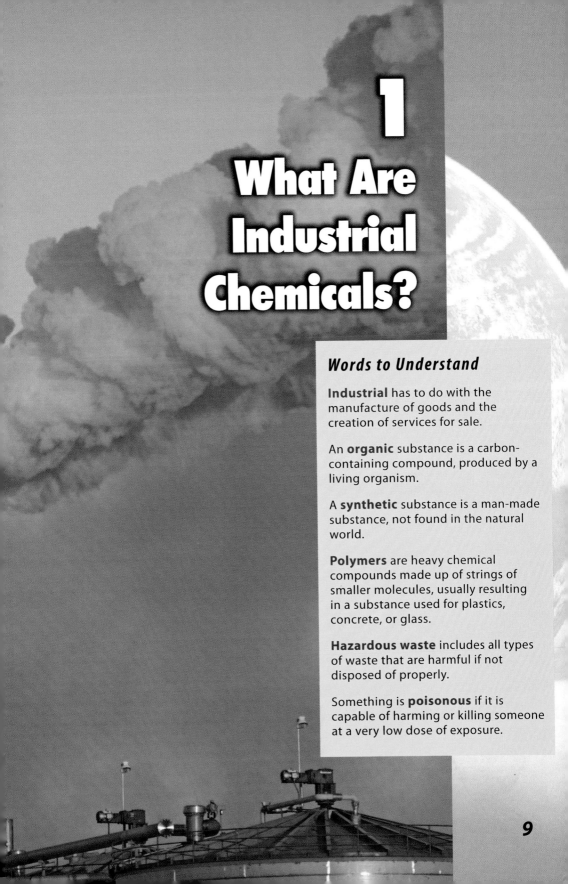

1 What Are Industrial Chemicals?

Words to Understand

Industrial has to do with the manufacture of goods and the creation of services for sale.

An **organic** substance is a carbon-containing compound, produced by a living organism.

A **synthetic** substance is a man-made substance, not found in the natural world.

Polymers are heavy chemical compounds made up of strings of smaller molecules, usually resulting in a substance used for plastics, concrete, or glass.

Hazardous waste includes all types of waste that are harmful if not disposed of properly.

Something is **poisonous** if it is capable of harming or killing someone at a very low dose of exposure.

The word "chemical" refers to a substance that is produced or used in a process involving changes to atoms or molecules. We sometimes think of all chemicals as bad or dangerous—but our entire world is made up of chemicals. For example, water—H_2O—and the oxygen in the air we breathe—O_2—are both chemicals.

What do you think of when you think of **industrial** chemicals? Do you imagine a factory tower with a plume of smoke coming out of it? Or do you picture black barrels of green slime glowing in the dark? Everyone seems to have a vague idea that industrial chemicals aren't good for us, but few know much more about them than that.

Today, industrial chemicals are all around us. Because there are an estimated 80,000 industrial chemicals currently in production, it would be difficult to give a single definition for them all. This book will focus on those industrial chemicals that are toxic to our health. What is toxic? Almost anything can be toxic if enough of it is introduced into the environment or into the body—even water is toxic to human beings in high quantities. In this book, a toxic substance is a chemical that enters the environment in a concentration high enough to have short-term or long-term effects that are harmful to that environment, or to animal and human life.

The chemical industry exists because not all chemicals occur naturally, as water does, or in such great quantities.

Sometimes human beings have to separate and create chemicals we need or desire. Take petroleum, for example. When **organic** substances—trees, animals, plants—die and decay, they leave behind elements (particularly carbon) that, under the right conditions, transform into coal and oil shale. As more time passes, these might further transform into a crude oil that human beings use to make many of the **synthetic** chemicals that exist today.

Not all chemicals are necessarily harmful. This is what the U.S. Environmental Protection Agency (EPA) has to say about it:

> Chemicals affect our everyday lives. They are used to produce almost everything we use, from paper and plastics to medicines and food to gasoline, steel, and electronic equipment. More than 70,000 chemicals are used regularly around the world. Some occur naturally in the earth or atmosphere, others are synthetic, or human-made. When we use and dispose of them properly, they may enhance our quality of life. But when we use or dispose of them improperly, they can have harmful effects on humans, plants, and animals.

A Brief History of Industrial Chemicals

The chemical industry has been around for a long time. Historians have traced this industry as far back as 7000 BCE, when Middle Eastern artisans refined alkali and limestone for the production of glass. Around 600 BCE the Phoenicians produced soap, and by the tenth century CE the Chinese had developed primitive gunpowder. Chemistry was quickly becoming a craft by the Middle Ages, when those practicing chemistry were known as "alchemists."

Alchemy is a word that most likely originated in Egypt and was brought home by European invaders. The earliest alchemists were Egyptian masters of goldwork. The word soon came to mean "the art of transformation"—turning one thing into another. Not long after, "alchemist" was changed to "chemist" in popular speech and the terms "chemistry" and "chemical" found widespread use.

By 1635, the pilgrims of the Massachusetts Bay Colony were using a basic chemical process to produce a salt for gunpowder and tanning animal hides. The first large-scale industry of chemicals began in the nineteenth century, with British entrepreneur James Muspratt's mass production of soda ash, a product needed for making soap and glass—two commodities in high demand at that time. Later, in the same century, more advances in chemistry would allow companies to develop dyes for the booming textile industry.

From this point on, the chemical industry began to grow at a rapid pace. Until World War I, Germany led the way in almost all sectors of the chemical industry. Their chemical processes were studied and envied by scientists around the world. In the 1890s, German companies began mass-producing sulfuric acid, caustic soda, and chlorine,

Alchemy was an early form of chemistry that combined magic, mysticism, and science to investigate the nature of various substances.

all three of which are still in use today. But it was during the beginning of the twentieth century, with the combination of World Wars and an increasing demand from the public, that the chemical industry really began to expand: plastics, fibers, synthetic rubber, and oil all came into mass production in the years between 1900 and 1950.

It was only in the 1950s that concern for toxic waste arose. It took another twenty years before the Environmental Protection Agency was created in 1970—years after the environment had already been significantly impacted by industrial chemicals. The EPA would gain more power over the years with the passing of the Toxic Substances Control Act (TSCA) in 1976, and the Food Quality Protection Act (FQPA) in 1996. The chemical industry as we know it today is vastly different than it was even as recently as a hundred years ago.

The Chemical Industry

The chemical industry is central to the modern world economy, converting raw materials (oil, natural gas, air, water, metals, minerals) into more than 70,000 different products. **Polymers** and plastics make up about 80 percent of the chemical industry's output worldwide. These chemicals are used to make various consumer products, and are essential to the agriculture, manufacturing, construction, and service industries. Major industrial customers include rubber and plastic products, textiles, apparel, petroleum refining, pulp and paper, and primary metals. The demand for industrial chemicals will continue to rise as the world continues to industrialize.

The European Union (EU) and the United States are the world's largest chemical producers, followed closely by Japan. The U.S. chemical output is $400 billion a year. The U.S. chemical industry records large trade surpluses and employs more than a million people in the United States alone. The chemical industry is also the second largest consumer of energy in manufacturing and spends over $5 billion annually on reducing pollution.

Did You Know?

If you were to add up profits from the chemical industries around the world, it would amount to approximately two trillion U.S. dollars a year!

The unavoidable truth is that our world is dependent on industrial chemicals. We would not be the society we are today without the chemicals produced by industry: pharmaceuticals (medicines), pigments for dyes in our clothing, monomers to make plastics, and precursors (chemicals that allow us to make other, more stable, products) are all possible because of the booming chemical industry. What might surprise many people, however, is that so many products we use every day are full of industrial chemicals whose harmful effects are not fully known: chemicals in the lotions we put on our skin, wrap our food in, paint on our fingernails, clean our kitchen stoves with, seal our floors with, and even put directly into our children's mouths.

Has the chemical industry crossed the line from creating a world helpful to all, to a world where chemicals go unchecked? For too long the chemical industry has operated under the idea that a chemical is innocent until proven guilty. But our environment and our health are not court cases, and there is an enormous body of evidence showing that toxic chemicals are everywhere in our world, posing a threat to everyone's health until they are fully tested. We call this a precautionary approach—minimizing our exposure to industrial chemicals expected to have toxic effects until the full range of their characteristics is understood.

What Makes a Chemical Dangerous?

When we dispose of these dangerous chemicals they become **hazardous waste**. Hazardous waste can come from any number of sources, but often it is the unwanted product of industrial processes. We can usually place these dangerous chemicals into one or more groups: corrosive, ignitable, reactive, or toxic.

Corrosive

A corrosive chemical wears away (corrodes) or destroys whatever it comes into contact with. Most acids are corrosives that can eat through metal, burn skin on contact, and give off vapors that burn the eyes.

Ignitable

Ignitable substances catch fire easily. They are fire hazards, can irritate the skin, eyes, and lungs, and sometimes give off harmful vapors.

Reactive

Reactive chemicals are those that explode or create **poisonous** gas when combined with other chemicals. Sometimes all that is needed for a reactive chemical to become deadly is air or even water. Some chemicals can react together to become deadly; for example, chlorine bleach and ammonia are reactive and create a poisonous gas when they combine with each other.

Petrochemical plants are responsible for releasing dangerous industrial chemicals into the environment.

Toxic

Toxic chemicals are poisonous or harmful to people and other life in some way. These chemicals are the main focus of this book. Toxic substances can be swallowed, absorbed through the skin, or breathed in the air.

Another factor in assessing a chemical's harmful effects is what we call the additive effect, or "cocktail effect." We use this term to describe the interaction of multiple toxic substances in our bodies and in our environment. The truth is we don't have the technology yet to properly understand how many different chemicals interact at once. We know that even though one chemical may seem harmless, once combined with other chemicals it can react and become a deadly chemical mixture, or cocktail.

Organic Chemicals

The term "organic chemicals" can be misleading because it makes one think of organic foods or other products—which generally means food or products made naturally,

Some industrial chemicals are produced as waste products of various manufacturing processes. Factories are then forced to deal with the question: Where can we throw away the barrels of toxic waste? There is literally no place these wastes can be put where they will not impact the environment and ultimately human life.

without chemicals. This, however, is not the case with organic chemicals, which are chemicals made artificially or naturally with compounds of carbon or hydrogen. In fact, organic chemistry is the foundation upon which hundreds of manmade substances are produced: plastics, solvents, flame retardants, preservatives, plasticizers, and many others are all the result of organic chemicals.

Key Traits of Toxic Chemicals

The first trait to look for in toxic chemicals is persistence. Persistent chemicals are chemicals that don't readily break down in the environment. Instead, they can remain intact in the environment and in organisms that consume them for many years, even decades. Because they don't break down, they continue to accumulate, posing a threat to our health.

A second trait of dangerous toxic chemicals is their lipophilic nature. Lipophilic literally means "lipid-loving" or "fat-loving"—chemicals that don't dissolve in water but stay in the fatty tissue of living organisms. If a chemical is both persistent and lipophilic, it can be taken in by organisms from contaminated environments and then accumulate along the food chain.

A third key trait of toxic chemicals is their tendency to bioaccumulate. Bioaccumulation is the process of accumulation along the food chain. Creatures lowest on the food chain consume chemically contaminated resources, then they in turn are consumed by creatures higher up on the food chain, which are consumed by other creatures higher up on the food chain, until they reach human beings. Along the way, the concentration of toxic chemi-

These plastic bowls look completely harmless—but they contain toxic chemicals that can leach into food.

cals accumulates until the concentration we consume is many times—sometimes millions of times—greater than that of the original environmental contamination.

POPs

Persistent organic pollutants (or POPs) are chemicals that possess all three of the key traits mentioned above. They persist for long periods of time in the environment, they build up in fatty tissue of living organisms, and they bioaccumulate. POPs are a large class of chemicals within the broad spectrum of toxic chemicals that exist today. These chemicals surround us and we interact with them directly or indirectly every day. Many have been banned already but persist in the environment in large quantities. Because many POPs are capable of traveling through water and air to regions far away from their original source, inhabitants of both rural and industrialized areas risk POP exposure.

Twelve of the most deadly POPs—called the "dirty dozen"—are already regulated by a United Nations treaty known as the Stockholm Convention, established in 2001. Commonly called the "POPs convention," this treaty established strict regulations for the following toxic chemicals:

1. PCBs
2. dioxins
3. furans
4. aldrin
5. dieldrin
6. DDT
7. endrin
8. chlordane

> **Ask the Doctor**
>
> **Q: What do I do if one of my friends or I spill toxic chemicals on our skin?**
>
> The first thing to do is remain calm. Then remove all contaminated clothing from your body. If you're helping someone, be sure to avoid contaminating your own skin. Then wash the exposed area with water for 10 minutes. Only after you've done this should you begin to wash the exposed area gently with soap and water. Once the situation is under control, immediately call 911 and give the emergency response operator the information he asks for.

What Are Industrial Chemicals? 19

Real People

Wilhelmina Scott lives in Bayview Hunters Point, a primarily low-income African American community within the city of San Francisco. The community also has the highest rate of breast cancer in the world. When Wilhelmina found a lump in her breast, she was almost certain she knew it was cancer, even before the doctor made an official diagnosis; after all, she'd supported her mother, two of her sisters, her best friend, and a neighbor as they battled breast cancer, so it was no surprise to her when she found out she too had the disease.

But Wilhelmina's daughter asked, "Why?" and started researching their community. She found out that Bayview Hunters Point has one the most polluting power plants in the United States. It also has a naval shipyard that dumped radioactive and other wastes into the ground and bay waters surrounding Hunters Point.

"They did this to us," Wilhelmina says. "So what do we do now? No more women should have to die. Someone needs to make this right."

9. hexachlorobenzene (HCB)
10. Mirex
11. toxaphene
12. heptachlor

Although only seven years have passed since the treaty was established, already many organizations such as the World Wildlife Fund (WWF) believe that many more chemicals with similar characteristics should be added to this list. Specifically, the WWF believes that certain brominated flame retardants (BFRs) should be monitored. BFRs are applied to a huge variety of products to prevent fire. Although they offer safety from fire to consumers, their toxic qualities should be weighed against their benefits, and a safer alternative sought out.

Types of Toxic Chemicals

Endocrine-Disrupting Chemicals or EDCs are chemicals that interfere with the endocrine or hormonal system. The endocrine system is an extremely complicated, intricate system of chemical messengers that regulate bodily functions such as metabolism, sexual development, and growth. The endocrine system is highly sensitive, and even the slightest adjustment in diet and lifestyle can have an effect. Endocrine-disrupting chemicals can significantly alter the natural course of hormones—especially at critical stages of development. EDCs can block natural hormone action, mimic it, or even have an opposing effect. The effects of EDCs were first seen in nature, where they interfered with animal behavior and sexual development. Evidence that EDCs affect human beings with similar problems—sexual development, birth defects, sperm count decline, and sexually related cancers—have all been suggested.

Leachates are chemicals that move from chemically treated products into the environment. Toxic chemicals can pose a threat if they do not remain in the product but leach out, or leave, and enter our households and bodies. They can accumulate in dust, air, and in food in chemically treated containers. A common example of a leachate is bisphenol A, a chemical used to line food cans that can enter the food they contain. Other examples are certain flame retardants that leach out of plastic casings and electrical equipment and enter the environment.

Volatile Organic Compounds (VOCs) are chemicals with the ability to evaporate at room temperature. These chemicals especially threaten us indoors. Paints, cleaning products, glues, PVC flooring, carpeting, and polishes all contain VOCs that may "off-gas" and fill a room with toxic chemicals that are dangerous to breathe. Off-gassing occurs when gaseous chemicals escape into the environment from substances to which they were applied.

What Do We Do with Toxic Chemicals?

The first priority is always to put them safely back into the systems in which they were created. This means reusing or recycling them. But if that isn't possible, factories will try to safely contain or dispose of these chemicals so that they aren't released into the environment where they can harm us. Over the past fifty years, technology has greatly helped us keep these chemicals out of the environment. Methods for storing and disposing of toxic chemicals include

The endocrine system's chemical messengers are easily thrown off from their normal functions. Endocrine-disrupting chemicals can significantly change the way these messengers (hormones) behave, especially at critical stages of development.

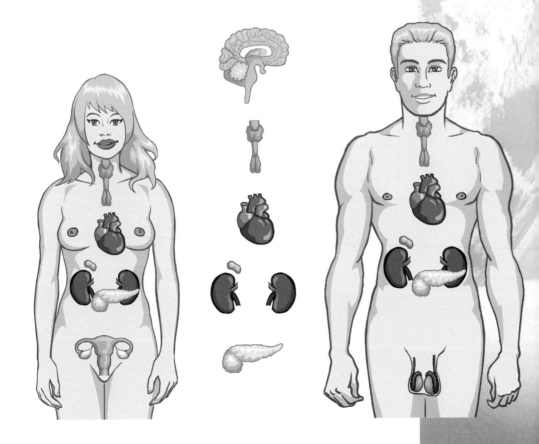

surface impoundments (storage in lined ponds), high-temperature incineration (controlled burning), secure chemical landfills (burying in the ground), and deep-well injection (pumping into underground wells).

How the Government Classifies Toxic Substances

The Agency for Toxic Substances and Disease Registry (ATSDR) lists forty-four chemical substances on its Web site as "toxic chemicals." This list literally covers substances from A to Z—from acetone to zinc. To better understand what makes a chemical toxic, let's take a closer look at this first toxic substance, acetone.

Landfills are not a perfect solution for dealing with industrial waste, since many of the chemicals will escape into the groundwater. This leachate treatment tank is an attempt to limit the toxic chemicals that escape from a landfill.

The ATSDR says that acetone is "a commonly manufactured chemical substance," which means that it's manmade in large quantities. But acetone also occurs naturally in the environment. If you were to study acetone in a laboratory it would look like a colorless liquid and have a distinct smell and taste. It evaporates easily, is flammable, and dissolves in water.

Acetone is used to make plastics, fibers, drugs, nail polish remover, and other chemicals. You can also find it in vehicle exhaust, tobacco smoke, and landfill sites. These are acetone's manmade uses. Interestingly, you can also find very small, harmless amounts of acetone in every human body, as acetone is also a product of the breakdown of body fat.

We know that in higher doses acetone becomes dangerous to human beings, but the ATSDR gives us little information about acetone beyond that. The symptoms of people exposed to various levels of acetone range from mild headaches and eye irritation to vomiting and unconsciousness. This report on acetone also gives us a long list of acetone's effects in animals. The animals were exposed to higher levels of acetone and the results were much more severe—from kidney disease to cataracts. But the ATSDR cannot say with certainty that these same results would occur in human beings exposed to high levels of acetone. The ATSDR also doesn't have information about acetone's ability to cause cancer. According to the report, "The EPA has determined that acetone is not classifiable as to its human carcinogenicity," which means the EPA doesn't know yet whether it can cause cancer.

Unfortunately, this report on acetone is characteristic of the U.S. government's knowledge of such toxic substances. Not enough funding exists to conduct proper testing on many substances suspected of being toxic, and without the full cooperation of the chemical industry, ordinary citizens will continue to be in the dark about which chemicals are harmful and which are not. In the meantime, these chemicals continue to hurt human beings and the environment.

Did You Know?

Journalist Nena Baker wrote a book called The Body Toxic: How the Hazardous Chemistry of Everyday Things Threatens Our Health and Well-Being. *In this book she states that more than 80,000 chemical substances are registered for commercial use in the United States. About 10,000 of these appear in products we use every day, but the EPA has no data on the potential human health effects of the vast majority of these chemicals.*

STRAIGHT FROM THE SOURCE

(From the 2008 Environmental Protection Agency (EPA) document Hazardous Substances and Hazardous Waste)

Hazardous Wastes

When hazardous wastes are released in the air, water, or on the land they can spread, contaminating even more of the environment and posing greater threats to our health. For example, when rain falls on soil at a waste site, it can carry hazardous waste deeper into the ground and the underlying groundwater. If a very small amount of a hazardous substance is released, it may become diluted to the point where it will not cause injury. A hazardous substance can cause injury or death to a person, plant, or animal if:

- A large amount is released at one time
- A small amount is released many times at the same place
- The substance does not become diluted
- The substance is very toxic (for example, arsenic).

Coming into contact with a substance is called an exposure. The effects of exposure depend on:

- How the substance is used and disposed of
- Who is exposed to it
- The concentration, or dose, of exposure
- How someone is exposed
- How long or how often someone is exposed.

Humans, plants, and animals can be exposed to hazardous substances through inhalation, ingestion, or dermal exposure.

- Inhalation—we can breathe vapors from hazardous liquids or even from contaminated water while taking a shower.
- Ingestion—we can eat fish, fruits and vegetables, or meat that has been contaminated through exposure to hazardous substances. Also, small children often eat soil or household materials that may be contaminated, such

as paint chips containing lead. Probably the most common type of exposure is drinking contaminated water.

- Dermal exposure—a substance can come into direct contact with and be absorbed by our skin.

Exposures can be either acute or chronic. An acute exposure is a single exposure to a hazardous substance for a short time. Health symptoms may appear immediately after exposure; for example, the death of a fly when covered with bug spray or a burn on your arm when exposed to a strong acid such as from a leaking battery.

Chronic exposure occurs over a much longer period of time, usually with repeated exposures in smaller amounts. For example, people who lived near Love Canal, a leaking hazardous waste dump, did not notice the health effects of their chronic exposure for several years. Chronic health effects are typically illnesses or injuries that take a long time to develop, such as cancer, liver failure, or slowed growth and development.

One reason chronic exposure to even tiny amounts of hazardous substances can lead to harm is bioaccumulation. Bioaccumulation occurs when substances are absorbed and stay in our bodies rather than being excreted. They accumulate and cause harm over time.

What Do You Think?

- Besides dosage, what are other factors of exposure?
- Which method of exposure—inhalation, ingestion, dermal exposure—do you think is most likely to occur in your household?
- Do you know anyone who has ever had an acute exposure to harmful chemicals?

Find Out More

What to Do in an Emergency Involving Toxic Chemicals
poisoncontrol.uchc.edu/emergency/index.htm

Fast Facts about Hazardous Substances/Waste
www.epa.gov/superfund/students/clas_act/haz-ed/ff_01.htm

A List of 302 Toxic Chemicals and their Profiles
www.atsdr.cdc.gov/toxpro2.html

Here's what you need to know

- Everyone is at risk from toxic chemicals. They are already in our environment, in the products we use, and in the food we eat.
- "Body burden" is a term given to the accumulation of toxic chemicals in our bodies over time.
- Different generations of people carry different amounts of chemicals in their bodies—younger generations generally have more.
- Dose and timing both play equal roles in how toxic chemicals affect our development.
- Extremely low doses of toxic chemicals can be harmful to children, infants, and fetuses whose stages of hormonal development are extremely vulnerable.
- Children and fetuses are fundamentally different from adults—not just in mass, but also in actual body composition. Because of this, they are exposed to toxic chemicals in far more dangerous ways.
- The unborn child receives toxic chemicals passed on by its mother.
- Both male and female reproductive organs are extremely vulnerable to the effects of toxic chemicals. Exposure to these chemicals during early development is particularly harmful, as studies done on children of DES mothers show.

Words to Understand

A **fetus** is an unborn child after the ninth week of development.

Embryonic is a term given to a child in its earliest stages of development in the womb—the first eight weeks of its life.

Incidence is the frequency of occurrence, or how often something happens.

The **placenta** is the organ formed in a mother's womb that protects and nourishes the child.

A child is in **utero** when it is still in its mother's body.

2
Who Is at Risk?

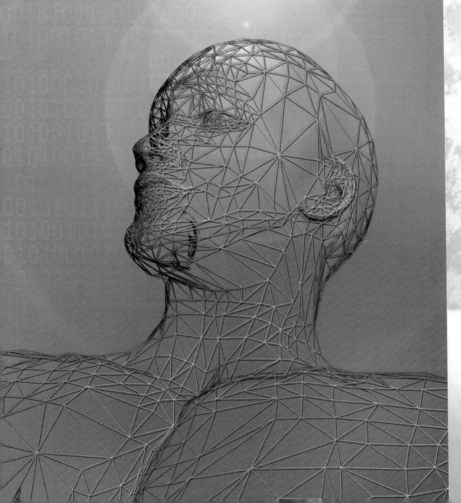

The short answer to this question is everyone. We are constantly being exposed to chemicals—in our environment, in the products we use, and in the food we eat. Every one of us, in fact, already carries a certain amount of toxic chemicals in his or her body.

Body Burden

"Body burden" refers to the amount of toxic chemicals in a body at a given time. How a chemical enters the body and how long it will remain there depends on the

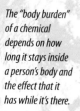

The "body burden" of a chemical depends on how long it stays inside a person's body and the effect that it has while it's there.

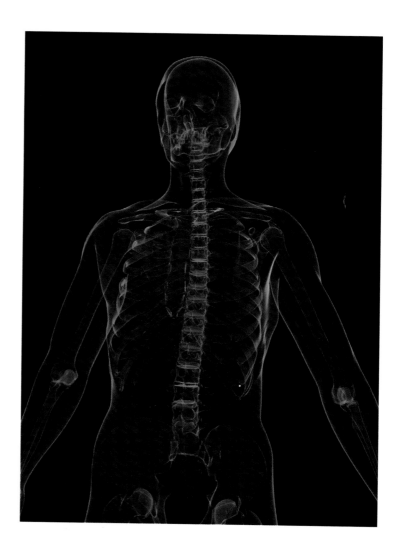

toxic chemical and the exposure to that chemical. Some chemicals are easily metabolized or broken down by our bodies. Others are much more persistent and will remain in our bodies for years, even decades. Even if the body is capable of metabolizing a chemical, if we are exposed to it daily, it will be an ongoing factor in our body burden. Because of industry and the nature of toxic chemicals, everyone has a body burden—from the hermit in his cave to the worker at a toxic-waste landfill. How is this possible? Consider the following: We are all contaminated from the beginning. A joint study conducted by the WWF and Greenpeace tested umbilical cords for the presence of thirty-five synthetic chemicals. The results showed that between five and fourteen of these chemicals were already in baby's bodies—even before they were born! Most of the chemicals found were from common household products.

The WWF conducted a cross-generational survey of body burden in 2004, comparing toxic body burdens of three generations of the same families. The survey revealed that children often have higher concentrations of toxic chemicals in their bodies, despite the fact they've had less time for exposure. The survey also showed that older generations tended to have more of the outdated chemicals in their blood, many of which have been banned by now, such as DDT and PCBs. The children, on the other hand, had elevated levels of more modern toxic chemicals like brominated flame retardants (BFRs) and perfluorinates—both of which will be discussed in detail later.

The important thing to know about body burden is that it is still being monitored, and unfortunately we have a long way to go before we know exactly which toxic chemicals we should avoid at all costs. Until then, be aware that your body already carries a variety of toxic chemicals, and until more is known, it's best to avoid them if possible. Some people face greater risks than others from toxic chemicals.

Who Is Most Sensitive to Toxic Chemicals?

A wide variety of chemicals are at work in our bodies, and it's not easy to assess the effects those chemicals have on our bodies. What has been easier to determine, however, is that certain stages of life are more vulnerable to harmful chemicals. **Fetuses** and children, for example, face particular challenges. Endocrine-disrupting chemicals (EDCs) also pose particular challenges to mothers and children. A central problem is that "safe" levels of exposure are generally determined based on tests done on full-grown adult males. These exposure levels must be adjusted for women, children, and infants.

In the sixteenth century, a Swiss physician named Paracelsus coined the phrase "the dose makes the poison."

Even before a fetus is born, it has industrial chemicals present in its body.

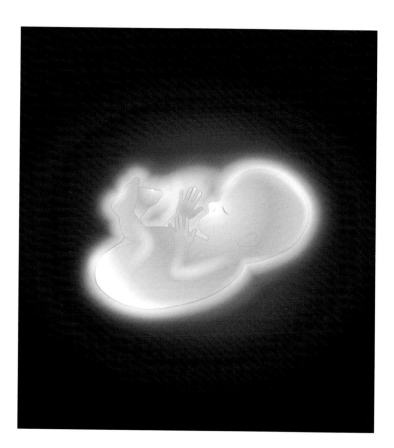

This concept—that the higher the amount of toxic chemicals, the more deadly they become—still holds true today, but with regard to toxic chemicals it should be reworded for accuracy: it is the dose plus the time of exposure that makes the poison. This is a huge shift in thinking, because it forces us to acknowledge that there are time periods when humans are particularly vulnerable, such as **embryonic** and fetal development.

Scientists have also discovered that some chemicals are actually more toxic in lower doses than in higher doses. What happens in these cases is that as the dose increases, a bodily reaction occurs and cancels out the negative effects. However, if the dose is so low that the body doesn't recognize it, it can go unnoticed and do harm.

These extremely low doses of toxic chemicals raise important questions when one considers the sensitivity of the endocrine system discussed in chapter 1. How much of a toxic chemical can really be considered "safe" when it comes to the hormones of our body?

Early Development

You might be surprised to know that the **incidence** of noninfectious diseases in children in the world is increasing. Allergies, asthma, diabetes, and various autoimmune disorders are all on the rise. In some countries, there are also rising rates of reproductive system defects such as lower sperm counts and earlier onset of puberty. The governments of the world can no longer afford to view children and infants as adults with less mass. The fact is, children are fundamentally different from adults because their bodies are still developing. Examine these factors:

- Infants and children eat more food than adults per pound of body weight. This means they take on more toxic chemicals for their smaller body weight.
- Once ingested, food stays in the child's digestive tract longer, leaving more time to absorb toxic chemicals. Take lead for example: a young child

Did You Know?

Almost 1 percent of Danish men are treated for testicular cancer, and a shocking 5.6 percent of Danish schoolboys have undescended testes. Early exposure to environmental factors may play a role in these statistics.

will absorb 50 percent of ingested lead, compared to 10 percent for adults.
- Babies and young children have a much higher intake of cow and breast milk, which leads to higher fat levels. Because many toxic chemicals are lipophilic, they seek out this fat and are stored at higher levels in children's bodies.
- Every human being has a blood-brain barrier that limits the ability of most toxic chemicals to enter the brain's lipid content, but in children this barrier is still in development. This means that there is an increased risk of exposing the brain to toxic chemicals.
- Children and infants breathe more rapidly, thus exposing them to higher levels of airborne toxic chemicals.
- Children and infants have higher dermal absorption, meaning their skin allows in more chemicals.
- Children and infants spend more time on the ground, increasing their exposure to chemicals in the house and in the soil.
- Children and infants frequently put things in their mouths.
- A variety of children's clothes are treated with toxic chemicals that, in combination with one another, could present a threat their health.

Fetal Sensitivity

When a child is still in the womb, extremely important stages of development are taking place all the time. The term "exquisite sensitivity" is given to fetuses during this time because of the very small amount of hormone that a fetus can react to—as tiny as one part per trillion. Any and all toxic chemicals that could affect a child's hormones need to be investigated.

One example of the power of EDCs in children is the disaster of diethylstilbestrol, or DES, a synthetic chemical given to pregnant women in the United States between the 1940s and 1970s to prevent miscarriages and other pregnancy complications. Unfortunately, little was known about the effects of such chemicals on fetuses, and DES led to many serious reproductive problems in both male and female offspring. Often these problems did not show themselves until the offspring reached puberty or adulthood, which serves as a warning about the long-term effects such chemicals can have.

Because of the widespread occurrence of toxic chemicals, every expectant mother will pass on some part of her body burden to her child, both through her **placenta** and through her breast milk. The placenta is designed to act as a protective barrier for the unborn child, but in the case of toxic chemicals it is rendered ineffective, as most toxic chemicals dissolve readily in fat and pass through to the child. Breast milk, too, is contaminated with toxic chemicals that can be passed to the child. But no one has suggested breast-feeding be stopped; breast milk is still the best source for children to receive nutrients they need. It is merely suggested that mothers make all possible attempts to reduce the amount of toxic chemicals in their own bodies.

Consider these additional facts about toxic chemicals and the unborn child:

- Even extremely low levels of chemical interference can devastate the fetus in its developmental stages.
- Toxic chemicals can cross the placenta and enter the fetus through its skin or digestive tract.
- The developing fetus lacks fat reserves, which act as a buffer to lipophilic chemicals in children and adults. Thus, these chemicals are more likely to end up in the few fat reserves a fetus has, such as in the brain.
- Infants born prematurely are particularly at risk of early exposure to phthalates (toxic chemicals

Toxic chemicals are particularly dangerous to pregnant woman, since these chemicals can alter the normal development of fetuses.

discussed in chapter 3) because of the PVC tubing they are hooked up to for medical care.

Male Reproductive Problems

Many toxic chemicals in circulation fall under the category of environmental estrogens. These chemicals mimic or block the hormone estrogen in our bodies and can cause significant harm to the sexual development of young men. Many environmental estrogens are chemicals in the ethoxylate family. They can be found in detergents, spermicides, paints, and some plastics. They are also used in pesticides and can enter the food chain through sprayed fruits and vegetables. The European Union has moved toward banning the use of ethoxylates, and Norway has already done so.

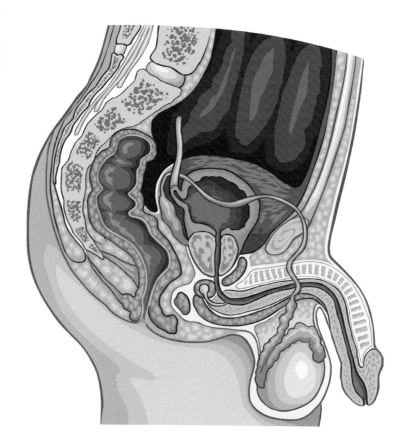

Chemicals called environmental estrogens can damage the male reproductive system.

A variety of male reproductive disorders are now being given the name testicular dysgenesis syndrome, or TDS. The TDS name is given to describe disorders that are either apparent at birth or show up in adulthood. They are important because they are thought to develop **in utero**. The development of the testes occurs almost entirely during early fetal development. Toxic chemicals that affect Sertoli cells—the cells responsible for producing sperm later in life—could be the source of various kinds of sexual disorders and testicular cancers in men.

The Price of Modern Life

Human bodies are paying the price for the conveniences offered by our modern world's industrial chemicals. Often these conveniences seem like necessities—but at what price to human life and health? Pesticides are examples of chemicals whose use seems more than justified; after all, insects and other pests interfere with the production of food, and chemicals that destroy these pests seem to make good sense. Unfortunately, however, pesticides carry with them heavy body burdens.

STRAIGHT FROM THE SOURCE

The Effects of DES

The U.S. Centers for Disease Control and Prevention gives this information on the known health effects of DES on daughters and sons:

Female Offspring:

Clear cell adenocarcinoma (CCA). A rare type of vaginal and cervical cancer. Approximately one in 1,000 DES daughters will be diagnosed with CCA. The risk is virtually non-existent among women not exposed to DES.

Reproductive tract structural differences. Including T-shaped uterus, hooded cervix, cervical cockscomb, and pseudopolyp.

Pregnancy complications. Ectopic (tubal) pregnancy and preterm (early) delivery.

Infertility. Difficulty becoming pregnant.

Male Offspring:

The most consistent research finding for DES sons indicates that they have an increased risk for noncancerous epididymal cysts, which are growths on the testicles. In one study, 21% of DES sons had noncancerous epididymal cysts, compared with 5% of unexposed men.

Other genital abnormalities. A few studies have reported that DES sons experience a greater likelihood of being born with undescended testicles (cryptorchidism), a misplaced opening of the penis (hypospadias), or a smaller than normal penis (microphallus). Because findings have been inconsistent, researchers cannot say with certainty that DES causes these types of genital abnormalities in DES-exposed men.

What Do You Think?

- If the CDC finds these studies inconsistent, why has DES been banned?

- Could toxic chemicals that affect Sertoli cells in men cause the types of conditions described in these studies?

Find Out More

To find out more about the effects of industrial chemicals on human bodies, check out these Web sites:

www.chem-tox.com/infertility/

www.wsws.org/articles/2005/dec2005/toxi-d07.shtml

www.bodyburden.org/

Here's what you need to know

- Pesticides are chemical substances used to harm or control unwanted organisms.
- Historically, insecticides have been used in successful civilizations for agriculture. More recently, insecticides reemerged to prevent disease during World War II, and afterward they were mass-produced for use in agribusiness all over the world.
- Organochlorines, one type of organic insecticides, persist, or remain, in the environment for very long periods of time.
- These persistent chemicals result in a process called bioaccumulation, which means even small amounts of toxins can accumulate and become harmful in larger quantities.
- DDT is one such persistent chemical. DDT and its harmful effects in nature brought international attention to the dangers of pesticides.
- Organophosphates are a less persistent form of organic pesticides, but they too pose dangers to human beings because they are more toxic in the case of acute exposure.
- Toxicologists and epidemiologists take different approaches to assessing pesticide risks—toxicologists use animal testing, while epidemiologists look at medical history.
- Scientists determine the NOEL for pesticides and other toxins in our food. NOEL stands for "no observable effects level."
- There are safer alternative pesticides currently being researched and developed to protect the public and environment from harm. These include pesticides designed to target specific pests without damaging others.
- Studies have found that certain groups of people, such as farmers, with more exposure to pesticides than average, are more susceptible to Parkinson's disease.

3
Pesticides

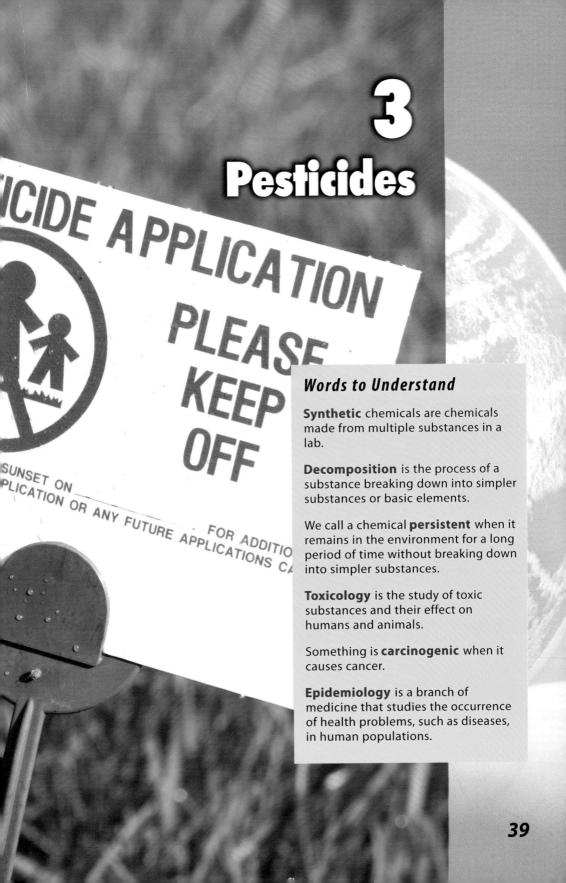

Words to Understand

Synthetic chemicals are chemicals made from multiple substances in a lab.

Decomposition is the process of a substance breaking down into simpler substances or basic elements.

We call a chemical **persistent** when it remains in the environment for a long period of time without breaking down into simpler substances.

Toxicology is the study of toxic substances and their effect on humans and animals.

Something is **carcinogenic** when it causes cancer.

Epidemiology is a branch of medicine that studies the occurrence of health problems, such as diseases, in human populations.

Did You Know?

One billion kilograms of pesticides are used every year in North America alone.

Pesticides are substances that kill or control an unwanted organism. All chemical pesticides inflict their damage by acting as toxins in the target organism, meaning they block a process vital to the organism and cause its death. The three types of pesticides most commonly used are insecticides, herbicides, and fungicides. Insecticides kill insects, herbicides kill plants, and fungicides kill fungi. The majority of pesticides used commercially are **synthetic** organic chemicals—combinations of organic (naturally occurring) substances.

Without pesticides, the world would be considerably different than it is today. Pesticides allow us to control the spread of insect-borne disease. They are also the reason we can produce as much food as we do. Half the pesticides in North America are used for agriculture, but that percentage increases globally: 85 percent of pesticides are used for agriculture in the worldwide community. In the United States, the majority of insecticides are used in the cotton industry, while the majority of herbicides are used in the growing of corn and soybeans. The type of pesticide used depends on the crop and the environmental factors of the country in which it's grown. Fungicides, for example, are used in large quantities in Colombia.

Pesticides and Health

Studies show that 80 to 90 percent of American homes have at least one synthetic pesticide. These include weed killers, algae controls for the pool, flea powders, and insect sprays for household bugs such as cockroaches. This means that inhabitants of almost every household are at risk for acute exposure to harmful pesticides.

In addition to the chemicals beneath sinks, half the foods eaten in the United States contain some level of pesticide. This has understandably caused alarm in the American public, and as a result many pesticides have been banned or restricted as the government and public have become increasingly aware. Yet debate continues. Some people defend synthetic pesticides by pointing out

Insecticides

that natural insecticides produced by plants themselves are already present in large amounts in our food.

Insecticides have been used for thousands of years in one form or another. As early as 1000 BCE, the Greeks burned sulfur to fumigate their homes. Fumigants are pesticides that kill an organism by inhalation. Insecticides have been extremely helpful to farmers and greater society, keeping insects away from crops and helping check the spread of disease. Few people realize how important insecticides

An apple grower sprays trees with pesticide that kills insects and fungi that could interfere with the apples' growth. Notice that this worker is wearing a mask to protect himself from the chemicals—but what will protect the consumers who eat these apples?

have been in reducing the incidence of diseases that have taken many lives in the past: malaria, yellow fever, bubonic plague, sleeping sickness, and (more recently) the West Nile virus. These diseases are all carried by insects or rodents, and insecticides form an essential line of defense against them.

Inorganic and metal-containing organic pesticides are usually very toxic to human beings and animals, especially in the high doses required to make them effective pesticides. Therefore, since the 1940s and 1950s the chemical industries of North America and western Europe have produced large quantities of organic insecticides. These organic insecticides are made largely from organochlorines, which are organic compounds that contain chlorine.

- Stable chemical structures that resist **decomposition**—they generally take a very long time to degrade back into the environment
- Low solubility in water—they don't dissolve easily, or at all
- High solubility in the fatty tissue of living organisms
- High toxicity in insects but low toxicity to humans

Hexachlorobenzene (HCB) is an example an organochlorine, an organic compound that contains chlorine, that has been a helpful protector of crops but a hazard to our health and environment. Organochlorines are **persistent** chemicals—chemicals that take long periods of time to dissolve into the environment.

HCB has been banned in most industrialized countries (although China, Mexico, and India have yet to decide), but it remains a widespread environmental contaminant. In addition, HCB has been linked to liver cancer in animal testing, and is speculated to do the same in human beings who are exposed to it over long periods of time. Currently, scientists estimate that 99 percent of North Americans have small amounts of the chemical in their body fat—a result of the process called bioaccumulation.

Bioaccumulation

Bioaccumulation happens in two phases: bioconcentration and biomagnification. Human beings can participate in both processes. Let's say a chemical pesticide such as HCB gets washed from the field into a nearby water supply. Because HCB is not very water soluble, it remains intact and is picked up by the gills of fish that breathe this contaminated water. The fish's gills, unlike the water itself, pick up and dissolve the molecules of HCB so that they become concentrated in the fish's fatty tissue. When an organism begins to build up larger and larger amounts of a chemical in its body, this is called bioconcentration. Next, a bigger fish eats many smaller fish with high amounts of HCB in their bodies, and soon it contains a large quantity itself. When a local fisherman hooks him and takes him home for dinner, he has no idea the fish he's about to eat has dangerously high levels of HCB in his body. This process is called biomagnification, and it is why even small amounts of toxic chemicals in our environment can accumulate and harm humans and animals.

DDT

Because of its significance in the media and culture of the twentieth century, DDT, or dichlorodiphenyltrichloroethane, requires our special attention. DDT is a persistent pesticide developed in 1939 by a Swiss chemist, Paul Muller, to combat clothes moths for the Swiss firm Geigy. During World War II, the allies adopted the pesticide and put it to use to protect citizens and troops. It was very effective against the mosquitoes that can carry malaria and yellow fever, against body lice that can transmit typhus, and against plague-carrying fleas. The World Health Organization (WHO) estimates that DDT may have saved the lives of five million people from malaria.

An ad from a 1947 issue of Time magazine promoted the use of DDT with these words: "The great expectations held for DDT have been realized. During 1946, exhaustive scientific tests have shown that, when properly used, DDT kills a host of destructive insect pests, and is a benefactor of all humanity."

DDT, THE "FAITHFUL ALLY OF THE GOOD HOUSEKEEPER"

In 1947, the United States Department of Agriculture distributed pamphlets encouraging the use of DDT. Here's what the USDA had to say to consumers:

> DDT was first used during the war for military needs by trained experts and under careful observation. It was used successfully to control the malaria-bearing mosquitoes, typhus-carrying lice, and other insects threatening the health of our Armed Forces. From this we know that DDT can be used safely. In the United States not a single case of DDT poisoning in humans has ever been proved when the material was used against insects.

DDT is a mild poison but it is safe when used according to these instructions:

> Take ordinary precautions in handling and storing DDT insecticides.
> Avoid applying it on eating utensils and food.
> Store it out of reach of children and where it will not be used by accident for flour, baking powder, or similar foods.
> Wash your hands when you have finished applying DDT.
> Do not spray oil solutions near open fires, because the oil may catch fire.

After WWII ended, DDT came into widespread use for public health issues, but in developing countries it was primarily used for agricultural purposes. Insects began to adapt and resist DDT, so farmers applied increasing amounts of DDT to their crops. In their book *Environmental Chemistry*, authors Colin Baird and Michael Cann finish the DDT story:

> Within the scientific community, reservations about DDT as the "perfect insecticide" began to be heard almost as soon as it first went into use. In particular, it was known that DDT in soil persisted for several years and could become concentrated in a food chain. The general public became aware of environmental problems associated with DDT upon the publication in 1962 of Rachel Carson's book *Silent Spring*. In it, she discussed the decline of the American robin in certain regions of the United States, due to its consumption of earthworms laden with the DDT used in massive amounts to combat Dutch elm disease. Carson's book stimulated widespread public concern about DDT and other pesticides. Through a series of legal hearings in the United States instigated by lawyers and scientists working with the Environmental Defense Fund, DDT was eventually banned or severely restricted in its use in most states. In 1973 the Environmental Protection Agency banned all DDT uses except those essential to public health. Similar bans were instituted by Sweden in 1969 and later in most other developed countries. DDT is still being used in developing countries to control disease or combat agricultural insects.

Organophosphates

Organophosphates are another kind of organic pesticide. They have an advantage over organochlorines because they don't accumulate in the environment and endanger us by chronic exposure. Unfortunately, they are generally more acutely toxic to humans. Those who apply them to crops and others who may be in close contact with them are at great risk. Exposure by inhalation, ingestion, or through the skin can lead to emergency health situations.

Organophosphates are particularly hazardous in developing countries, where there is a lack of information about their deadly nature. This lack of information, in addition to a lack of money for protective gear, has led to the death of many agricultural workers. In Ecuador, for instance, a potent organophosphate is used to keep pests away from the potato crop, and does so effectively, but at the expense of some potato farmers' lives. The United Nations and the World Health Organization estimate the number of persons who suffer illness from acute pesticide exposure is in the millions annually. Each year, 10,000 to 40,000 people die from poisoning, and about three-quarters of them are in developing countries. In the United States, about 20,000 people are hospitalized annually for poisoning from pesticides, and about thirty people die from it each year.

Two Methods for Assessing Risk

Toxicology determines the safety of each type of pesticide using tests done on animals. Animals are exposed to increasing doses of pesticide until effects are seen. These effects are recorded and a no observable effects level (NOEL) is determined. This level can be deceiving, especially if not enough animals were tested and effects at low levels were not uncovered. Also, most toxicologists agree that **carcinogenic** effects cannot be detected using this method, as cancerous results take much longer to reveal themselves.

Epidemiology is interested in these long-term effects of pesticides and other toxins. In epidemiology, rather than running tests on animals, scientists examine the health history of a select group of human beings. They examine their current ailment—often cancer—and try to relate it to differences in substances to which they have been exposed in the past.

In order to make toxicology data useful, we translate the NOEL for animal test subjects to a NOEL for human beings. This is done by dividing the NOEL from animal studies by a safety factor, usually 100. The resulting value is called the maximum acceptable daily intake (ADI), or maximum daily dose. As you can imagine, this value is far from exact, and by using this process we run the risk of arriving at a value too high or too low. Because of this uncertainty, some scientists suggest we divide the NOEL by a further factor of 10 to protect children, whose bodies are more vulnerable to toxins. In agreement with such suggestions, the 1996 Food Quality Protection Act in the United States required that the EPA set limits on pesticide levels in foods 10 times lower than what is considered safe for adults.

Alternative Insecticides

Most of the insecticides discussed thus far are known as broad-spectrum insecticides because they target enzymes common to most insects—and to humans. This lack of specificity makes them dangerous both to human beings and to beneficial insects, such as bees or ladybugs. The response has been to develop insecticides that target only one type of insect.

Chemists accomplished this by developing insecticides that interrupt a biological function unique to the targeted pest. These pesticides are generally safer for the environment and human beings and thus are classified as "reduced-risk pesticides" by the EPA. To qualify for this classification, a pesticide must meet the following requirements:

- It reduces pesticide risks to human health.
- It reduces pesticide risks to nontarget organisms (bees, for example).
- It reduces the potential for contamination of valued environmental resources.

Did You Know?

According to California's Department of Pesticide Regulation (DPR): There are more than 865 active ingredients registered as pesticides, which are formulated into thousands of pesticide products available in the marketplace. About 350 pesticides are used on the foods we eat and to protect our homes and pets.

Pesticides and Parkinson's Disease

Because so many pesticides are currently in use, and because they've been consistently used for over fifty years, it is difficult to definitively link a particular pesticide to a particular disease.

That being said, plenty of evidence indicates that those most often exposed to pesticides (usually farmers) are at higher risk for Parkinson's disease (PD). Researchers at the Harvard School of Public Health found that people exposed to chronic, low-dose amounts of pesticides had a 70 percent higher incidence of PD than those not exposed. Future studies are needed to determine which type of pesticide is connected to PD, but the results remain significant.

Justifying Chemicals

Pesticides protect us from disease and increase food production, two very worthy goals. Meanwhile, other chemicals are the byproducts of equally worthy goals. We take for granted the presence of these various products in our lives—but if we look a little deeper, we can begin to count the cost of these chemicals' use. Does the end product justify the use of an industrial chemical?

Ask the Doctor

Q: Do certain foods contain more pesticide residue than others?

Yes. Fruit sometimes contains more pesticide residue than other foods because fruits are particularly prone to damage from pests and the use of pesticides may prevent this damage. This use is taken into account when pesticides are authorized and trials are undertaken to ensure that pesticide residues left in the crop would not harm anyone, including children. Just peeling most fruits dramatically reduces the amount of pesticide exposure.

Real People

In Bhopal, India, 1984, a pesticide manufacturing plant called Union Carbide accidentally leaked 40 tons of methyl isocyanate (MCI) gas into the surrounding city. Nearly four thousand people would die that night from poisoning, but the long-term effects of the disaster are still being felt today. The Bhopal Disaster, as its known, is considered one of the worst industrial disasters in history. Taken from www.bhopal.org/whathappened.html, here are a few first-person accounts of people who survived that horrendous night:

"At about 12.30 am I woke to the sound of my baby coughing badly. In the half light I saw that the room was filled with a white cloud. I heard a lot of people shouting. They were shouting 'run, run.' Then I started coughing with each breath seeming as if I was breathing in fire. My eyes were burning."

"It felt like somebody had filled our bodies up with red chillies, our eyes tears coming out, noses were watering, we had froth in our mouths. The coughing was so bad that people were writhing in pain. Some people just got up and ran in whatever they were wearing or even if they were wearing nothing at all. Somebody was running this way and somebody was running that way, some people were just running in their underclothes. People were only concerned as to how they would save their lives so they just ran."

"The poison cloud was so dense and searing that people were reduced to near blindness. As they gasped for breath its effects grew ever more suffocating. The gases burned the tissues of their eyes and lungs and attacked their nervous systems. People lost control of their bodies."

STRAIGHT FROM THE SOURCE

Canada and Pesticides

The agency that regulates pesticides in Canada is known as the Pest Management Regulatory Agency (PMRA). The following information comes from its Web site, pestcontrolcanada.com:

A pesticide manufacturer who wishes to sell a pesticide in Canada must first register that product under the Pest Control Products Act and follow the registration process managed by the PMRA. The pesticide is put through a series of detailed scientific tests and studies, which provides answers to the following questions:

- Where, how and by whom will the pesticide be used?

- What is its toxicity level?

- Are there potential health hazards to users and/or bystanders?

- Will food and drinking water be affected?

- What are the short-term and long-term impacts on the environment?

What Do You Think?

- Are there any criteria that the PMRA has forgotten to include in this list?

- Would bioaccumulation be considered a short-term or long-term impact on the environment?

Find Out More

Learn more about pesticides at these Web sites:

www.pestcontrolcanada.com

www.epa.gov/pesticides/health/human.htm

www.who.int/topics/pesticides/en/

Here's what you need to know

- Nonpesticide toxic chemicals is a category that includes dioxins, PCBs, and phthalates.
- Some research has linked liver damage and skin rashes to acute exposure to large amounts of dioxins. TCDD is the most dangerous dioxin. It acts as a carcinogen in animals and is suspected to be a carcinogen in humans as well.
- Agent Orange was a military campaign that exposed thousands of people to dioxins in large amounts. The health effects were various, and are still being felt today.
- The Seveso Disaster in Italy is another example of dioxins' harmfulness in large quantities. Those affected showed an increased risk of breast cancer.
- PCBs are synthetic compounds mostly found in plastics and electrical insulators. They were banned in 1977 but are persistent, and thus remain throughout the environment.
- Phthalates are chemical compounds most commonly found in soft plastics.
- Speculation about phthalates' harmful effects in adults have not been confirmed, but steps have already been taken to protect children from phthalates found in plastic toys.

Words to Understand

Byproducts are substances produced during the making of something else.

Chloracne is an acne-like skin disorder caused by exposure to some chemicals.

Defoliants are chemicals sprayed onto plants to cause the leaves to fall off.

Congeners are substances that all have similar characteristics.

Physical **abnormalities** are occurrences in the body that are not normal.

4
Dioxins, PCBs, and Phthalates

The chemicals discussed in this chapter are not pesticides, but like pesticides they too are largely the result of industrial processes. In some cases they are merely **byproducts** of natural processes. In other cases they are the direct result of commercial enterprises, such as the plastics industry. Three categories of chemicals are the most prevalent in our environment—dioxins, PCBs, and phthalates—and these are the ones that require our society's immediate attention.

Dioxins

Dioxins are a group of chemicals that occur as byproducts in the manufacture of organochlorides (discussed in the previous chapter), in the incineration of plastics,

You probably take for granted that paper is white— but just how important is it to you? Paper is white because of dioxins, which are used to bleach it.

in the bleaching of paper, and also as a result of natural processes, such as forest fires and volcanoes. Before the Industrial Revolution, dioxin levels were approximately three times lower than they are today, meaning human industry is largely responsible for current levels of dioxins. Small amounts of dioxins can be found in almost every human being, but those with the highest levels tend to live near industrialized cities.

According to the most recent U.S. EPA data, the major sources for dioxins are:

- coal-fired utilities
- waste incinerators
- metal smelting
- diesel trucks
- land pollution by sewage sludge
- burning treated wood
- trash burn barrels

In total, these sources account for 80 percent of total dioxin emissions. Dioxins are also present in small amounts in materials used by humans. Plastics, resins, bleaches, and an assortment of other products contain small amounts of dioxins. Most scientists agree that there is little to no proof that small amounts of dioxin are harmful, but research into the long-term effects of small-dose dioxin exposure is still being conducted.

Short-term exposure of humans to high levels of dioxins may result in skin lesions, such as **chloracne** and patchy darkening of the skin, and altered liver function. Long-term exposure is linked to impairment of the immune system, the developing nervous system, the endocrine system, and reproductive functions. Chronic exposure of animals to dioxins has resulted in several types of cancer. TCDD is the most toxic of dioxins. It was evaluated by the WHO's International Agency for Research on Cancer (IARC) in 1997. Based on animal data and on human epidemiology data, TCDD was classified by IARC as a "known human carcinogen." However, TCDD does not affect genetic material and there is a level of exposure below which cancer risk would be highly unlikely.

Because dioxins are everywhere, almost all of us carry certain levels of them in our bodies. These small levels of chemicals persist in us and accumulate in our bodies over years. Because of the sheer amount of variety with which these chemicals exist in our bodies, it is difficult to say how they affect our health. But they have been blamed for a broad spectrum of ailments, from high blood pressure to dysfunctional immune systems. Although most scientific studies require more research before conclusions are drawn, it is clear already that reducing the amount of chemicals in our environment, and therefore in our diet, is of benefit to everyone.

Agent Orange

Plenty of information, however, indicates that intense, short-term exposure to dioxin can be devastating. Perhaps the most famous example of dioxin contamination is the U.S. military campaign during the Vietnam War called Agent Orange. The primary goal of Agent Orange was to spray Vietnam's forests with **defoliants** in order to uncover enemy forces in hiding. The defoliant's byproduct was a dioxin, and both the defoliant and dioxin poisoned the soil and the Vietnamese people for generations to come. Dioxins are extremely persistent, and remain in the environment, accumulating in plant and animal life and threatening human beings.

The number of reported developmental disorders and mutations resulting from Agent Orange is enormous. Those most affected were U.S soldiers who directly handled Agent Orange and Vietnamese children exposed to high quantities of the deadly herbicide and dioxin. In addition, the U.S. Department of Veterans Affairs has listed the following conditions as side effects in children of veterans exposed to Agent Orange: prostate cancer, respiratory cancers, multiple myeloma, type II diabetes, Hodgkin's disease, non-Hodgkin's lymphoma, soft tissue sarcoma, chloracne, porphyria cutanea tarda, peripheral neuropathy, and spina bifida.

Seveso Disaster

Another health disaster concerning dioxins occurred in Seveso, Italy, in 1976, at a small chemical manufacturing plant. The company, called ICMESA, created quantities of herbicide for commercial sale. One day in July, workers left the building without fully halting all chemical reactions. The chemical process continued unchecked while heat built up, which led to an explosion. This explosion ignited the herbicide and released dioxins into the surrounding environment. A large number of animal deaths resulted: 3,300 small animals, mostly poultry and rabbits, were found dead within days, and approximately 80,000 animals had to be slaughtered in order to prevent contamination of the food chain. Over a thousand people

Vietnamese young people with disabilities reveal just one of the consequences of exposure to this dangerous form of dioxin.

were directly affected, with 447 reporting skin lesions as a result of the concentration of dioxins in the air. No serious health effects were found for many years. However, recent studies reveal that several types of cancer have appeared in those most exposed to dioxins from the explosion. Breast cancer, in particular, was found to have increased among women whose blood samples were taken soon after the disaster.

PCBs

PCBs (polychlorinated biphenyl) are synthetic chemicals made from the combination of over 209 different chlorinated compounds, or **congeners**. PCBs do not occur naturally but were produced synthetically until 1977, when they were banned in most countries. Since the 1950s, over one million metric tons of PCBs have been produced—about half in the United States, and the rest divided between France, Japan, and the former Soviet bloc. PCBs are either oily liquids or solids with a clear or tinted-yellow appearance. They were widely used in plastics and as insulation for electrical equipment, as they don't burn easily. Many electrical devices from before 1977 contain large amounts of PCB in their components.

The Agency for Toxic Substances and Disease Regulation (ATSDR) breaks down how PCBs entered the environment:

- PCBs entered the air, water, and soil during their manufacture, use, and disposal; from accidental spills and leaks during their transport; and from leaks or fires in products containing PCBs.
- PCBs can still be released to the environment from hazardous waste sites; illegal or improper disposal of industrial wastes and consumer products; leaks from old electrical transformers containing PCBs; and burning of some wastes in incinerators.
- PCBs do not readily break down in the environment and thus may remain there for very long

periods of time. PCBs can travel long distances in the air and be deposited in areas far away from where they were released. In water, a small amount of PCBs may remain dissolved, but most stick to organic particles and bottom sediments. PCBs also bind strongly to soil.
- PCBs are taken up by small organisms and fish in water. Other animals that eat these aquatic animals as food also take in PCBs. PCBs accumulate in fish and marine mammals, reaching levels that may be many thousands of times higher than in water.

Studies have examined the health effects of exposure to large amounts of PCBs. The most common effects observed in these studies were skin conditions such as rashes and acne. Some people showed signs of liver damage. Animals given large amounts of PCB for short periods of time showed signs of liver damage and some died. Animals that ate smaller amounts of PCBs in food over several weeks or months developed various kinds of adverse health effects, including anemia; acne-like skin conditions; and liver, stomach, and thyroid gland injuries. Other effects of PCBs in animals include changes in the immune system, behavioral alterations, and impaired reproduction.

PCBs are thought to be carcinogenic, based on animal testing in which rats developed cancer after two years of exposure to high doses of PCBs. The Department of Health and Human Services (DHHS) has concluded that PCBs may reasonably be anticipated to be carcinogens. The EPA and the International Agency for Research on Cancer (IARC) have also determined that PCBs are probably carcinogenic to humans.

Researchers have found that women exposed to large amounts of PCBs in the workplace or women who ate large amounts of PCB-contaminated fish had babies that weighed less than average. Some babies born to mothers who ate PCB-contaminated fish exhibited behaviorial problems, such as decreased motor skills and short-term

Real People

Hong Hanh is falling to pieces. She has been poisoned by the most toxic molecule known to science; it was sprayed during a prolonged military campaign. Hong Hanh's story, and that of many more like her, is quietly unfolding in Vietnam today. Her declining half-life is spent unseen, in her home, an unremarkable concrete box in Ho Chi Minh City, filled with photographs, family plaques and yellow enamel stars, a place where the best is made of the worst.

Hong Hanh is both surprising and terrifying. Here is a 19-year-old who lives in a 10-year-old's body. She clatters around with disjointed spidery strides which leave her soaked in sweat. When she cannot stop crying, soothing creams and iodine are rubbed into her back, which is a lunar collage of septic blisters and scabs. "My daughter is dying," her mother says. "My youngest daughter is 11 and she has the same symptoms. What should we do? Their fingers and toes stick together before they drop off. Their hands wear down to stumps. Every day they lose a little more skin. And this is not leprosy. The doctors say it is connected to American chemical weapons we were exposed to during the Vietnam war."

There are an estimated 650,000 like Hong Hanh in Vietnam, suffering from an array of baffling chronic conditions. Another 500,000 have already died. The thread that weaves through all their case histories is defoliants deployed by the US military during the war.

(From an article by Cathy Scott-Clark and Adrian Levy printed in the March 29, 2003 issue of The Guardian.)

memory. Other studies suggest immune system problems in children born to mothers exposed to large amounts of PCBs. Although infants are most likely exposed through breast milk, none of these studies prove that the negative effects of PCBs outweigh the benefits of breast-feeding.

Both in Japan in 1968 and in Taiwan in 1979, many people unintentionally consumed PCBs that had accidentally been mixed with cooking oil. In both instances, the PCBs were heated and unknowingly cooked into food. Both of these tragic cases provided valuable information about the acute effects of PCB poisoning. Brain development was measured by IQ scores given to children of Taiwanese mothers most exposed to PCBs. Scores were much lower for children born to mothers after exposure to PCB. Children whose fathers were exposed to PCB showed no harmful effects.

> **Ask the Doctor**
>
> **Q: How can I protect my little brother and the rest of my family from PCBs already in the environment?**
>
> First of all, you can be careful to follow any warnings that the government gives about areas contaminated with PCBs. Be sure not to eat fish from lakes that are known to have high levels of PCB. Also, don't let your little brother play with old appliances, electrical equipment, or transformers, since they may contain PCBs. Take note of any areas in your town where electrical fires or hazardous waste occur, and keep your brother away from their soil, because children have a tendency to put dirty hands and toys in their mouths.

Phthalates

Phthalates are a type of chemical added to numerous consumer products. In 1994, almost all phthalates (close to 87 percent) in the United States were used as plasticizers, or softening agents, in vinyl products. Phthalates compose a large percentage of soft vinyl products; they are responsible for more than 40 percent of an average soft vinyl product's weight.

Phthalates are everywhere in modern society, which means human beings are exposed to them every day. The Organization for Chemical Body Burden gives us this information:

> Humans are widely exposed to phthalates because vinyl is an everyday plastic used to make anything

from home furnishings (flooring, wallpaper), medical devices (catheters and IV and blood bags), and children's items (infant feeding bottles, squeeze toys, changing mats, teethers), to packaging (disposable bottles, food wrap). Beyond vinyl, humans are further exposed to phthalates in cosmetics and scented products such as perfumes, soaps, lotions, and shampoos. Phthalates are also added to insecticides, adhesives, sealants and car-care products.

Because we are exposed to phthalates so often, significant research has been done to determine what effect, if any, they have on our health. As is often the case, phthalates remain widespread because there has not been sufficient evidence to link disease to exposure to small amounts of phthalates.

However, in 2005 the European Parliament voted to ban the use of six phthalates in children's toy manufacturing. A study done at the University of Rochester in New York linked toys made with phthalates to a higher risk of genital **abnormalities** in baby boys. Because children chew on these toys and because they have a significantly lower body mass, they are at greater risk for harm.

The National Toxicology Program (NTP) continues to monitor pregnant women exposed to high amounts of phthalates in order to understand how child development is affected by phthalates. The EPA has classified phthalates as a "probable human carcinogen." Animal studies reveal that rats and mice fed phthalates show an increase in liver cancers. In addition, animals fed phthalates produced offspring with abnormal patterns of sexual development; in some cases, rats fed high doses produced offspring with an extra rib. Unfortunately, not enough evidence exists yet to link these risks to human beings.

And the List Goes On

The problems presented by industrial chemicals are hard to tackle, partly because these chemicals are so woven through our daily lives. What's more, the list of toxic chemicals is not a short one. Unfortunately, pesticides, dioxin, PCBs, and phthalates are not the only chemicals we need to worry about.

STRAIGHT FROM THE SOURCE

Dioxins Around the World

The World Health Organization (WHO) gives us this information about dioxins in the international community:

Some dioxin contamination events have been more significant, with broader implications in many countries.

In July 2007, the European Commission issued a health warning to its Member States after high levels of dioxins were detected in a food additive—guar gum—used as thickener in small quantities in meat, dairy, dessert or delicatessen products. The source was traced to guar gum from India that was contaminated with pentachlorophenol (PCP), a pesticide no longer in use. PCP contains dioxins as contamination.

In 1999, high levels of dioxins were found in poultry and eggs from Belgium. Subsequently, dioxin-contaminated animal-based foods (poultry, eggs, pork) were detected in several other countries. The cause was traced to animal feed contaminated with illegally disposed PCB-based waste industrial oil.

In March 1998, high levels of dioxins in milk sold in Germany were traced to citrus pulp pellets used as animal feed exported from Brazil. The investigation resulted in a ban on all citrus pulp imports to the EU from Brazil.

Another case of dioxin contamination of food occurred in the United States of America in 1997. Chickens, eggs, and catfish were contaminated with dioxins when a tainted ingredient (bentonite clay, sometimes called "ball clay") was used in the manufacture of animal feed. The contaminated clay was traced to a bentonite mine. As there was no evidence that hazardous waste was buried at the mine, investigators speculate that the source of dioxins may be natural, perhaps due to a prehistoric forest fire.

What Do You Think?

- Is illegal dumping of hazardous waste individual nations' problems? Or is it a global problem?

- Are dioxins always the result of manmade processes?

- Are there bodies of water near you that have been tested for dioxins? If so, has your community continued to fish in those bodies of water?

Find Out More

www.chemicalbodyburden.org

www.who.int/mediacentre/factsheets/en/

www.atsdr.cdc.gov/

Here's what you need to know

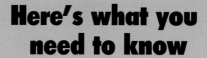

- There are hundreds of toxic chemicals in consumer products we use every day, but this chapter will focus only on a few that are more common than others.
- BPA is a synthetic estrogen and the main ingredient in polycarbonate plastic.
- BPA is very persistent and is found throughout the environment. Children and infants are especially vulnerable to BPA's effects.
- BFRs are found in many consumer products; they are synthetic chemicals added to products to make them flame resistant.
- BFRs are known to mimic thyroid hormones and can disrupt brain function as well.
- Safer alternatives to BFRs are available if one seeks out furniture and paneling not made specifically for indoor use.
- PCFs are applied to clothing and cookware for their nonstick, heat, and odor-resistant qualities.
- PCFs can vaporize if temperatures are high enough, releasing toxic gases into the air.
- Alternatives to PCFs include a variety of porcelain and metal cookware.
- Formaldehyde is a very common chemical in household cleaners and treated wood products.
- Formaldehyde has a tendency to vaporize and fill closed indoor areas with toxic gas.

5 Common Toxic Chemicals and Their Health Effects

Words to Understand

A **neurotoxic** chemical is a chemical that damages the brain.

Impervious means incapable of being affected.

Did You Know?

Almost three million tons of BPA are produced annually worldwide.

Did You Know?

Generally, women have higher amounts of BPA in their bodies than men do. This may be due to exposure, but is probably due to differences in how the sexes break down this chemical.

Did You Know?

BPA is so pervasive that it has even been found in deep-water whales.

Our world is full of harmful chemicals. This chapter will discuss just a few.

Bisphenol A (BPA) or Polycarbonate

BPA had its beginnings as a synthetic estrogen, or man-made hormone, but quickly found an indefinite use as the main chemical used to create polycarbonate. Polycarbonate is an amazing plastic that resists heat and odors, can be nearly shatterproof, and is of very low weight. BPA seems like a miracle plastic, but it's not: it acts as a synthetic estrogen and therefore has the potential to affect a human being's hormones. BPA is especially risky because it is used in the packing of so many foods, leaving the possibility for BPA to leach into and contaminate the foods we eat.

BPA is invaluable to the chemical industry because of its many uses. BPA is in countless modern products: baby bottles, cell phones, DVDs, plastic lunch boxes, eyeglass lenses, linings of cans—everything from paints to dental sealants. Its usefulness is one reason why regulation has emerged so slowly, but it should also be noted that BPA was introduced as a pharmaceutical, not as a polycarbonate. Because of this, it was considered more helpful than harmful, and proper regulation has never been established for BPA.

Until the recent controversy over the harmful effects of BPA, thousands of baby bottles contained this harmful chemical. Many have been recalled, but some still contain BPA.

Safe Alternatives

The best way to avoid BPA is to avoid polycarbonates. This means replacing polycarbonate plastics whenever possible—or, where they are necessary, be sure to discard these plastics after a few months. For instance, mothers should use glass instead of plastic baby bottles. Cut down on the amount of canned products you use, especially if the product they contain is high in fat content. Avoid microwaving food in plastic containers.

BPA is found in many plastics—water bottles, plastic eating utensils, and plastic food containers. Heating these containers in microwaves may increase the chance of BPA leaching out of the product and into foods. Most exposure to BPA is by ingestion. BPA is often in contact with food and drink, and can leach into these sources. BPA can be transferred from mother to child through placenta or breast milk, and BPA can accumulate in house dust and environmental soil and be inhaled.

Did You Know?

BPA affects reproductive hormones in snails, in some cases causing them to produce so many eggs that they burst.

Brominated Flame Retardants (BFRs)

BFRs are synthetic chemicals added to many consumer goods to prevent them from burning too quickly. They are added to a large number of products, including furnishings, carpeting, bedding, children's clothing, and electrical goods. BFRs exhibit all three properties of many hazardous chemicals in that they are persistent, bioaccumulative, and toxic. They have been shown to cause negative health effects in animals; the liver, brain, and nervous system were all harmed in animal studies.

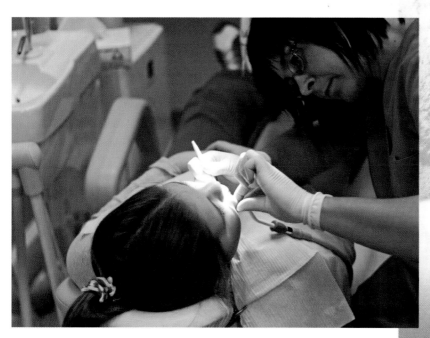

BPA is used for dental sealants to prevent tooth decay, especially in children.

Did You Know?

BPA has definite effects in the endocrine system of animals, and very likely affects human hormones as well.

Did You Know?

Evidence suggests that very low doses of BPA can affect fetal growth.

BFRs are injected into products we use every day and are a cause for alarm because they are in close contact with our environment and us. Products that contain the highest levels of BFRs are soft furnishings, the plastic casings of small electrical goods such as computers and televisions, and carpeting and floorings with synthetic fibers. Sometimes as much as 10 percent of these products is made of BFR. Obviously, flame retardants can save lives and should not be thrown out entirely. But it is important to ask whether or not safer alternatives should be sought out before toxic chemicals are sprayed onto products we use every day.

The BFRs of greatest concern to human beings are the PBDEs (polybrominated diphenylethers). The European Union has banned two types of PBDEs and a third PBDE is currently under investigation to determine its harmfulness. Because BFRs are so persistent and are being found in the environment and in human beings in increasing amounts, many have suggested they be added to the list of chemicals covered by the Stockholm Convention.

A very serious hazard related to BFRs is their structural similarity to thyroid hormone. BFRs are suspected

Safe Alternatives

Fortunately, much can be done to avoid BFRs if one seeks out the right manufacturers. Many companies are anticipating a possible future ban on all PBDEs, and therefore are working to replace BFRs with safer alternatives. Some of these companies include Dell, Ericsson, Philips, and NEC. The extremely popular Swedish furniture store IKEA has already phased out BFRs from its product line.

Actively replace products in your home treated with BFRs. This may mean major changes, such as replacing carpet with wood flooring. Or the changes you choose could be smaller, such as choosing metal casing for your computer instead of a plastic case treated with BFR.

Furniture with fewer or no BFRs can be found as well. Usually antiques or locally made furniture will not have been treated with BFRs.

to affect thyroid and hormone balance and may cause harm to the brain as well. BFRs, like many other toxic chemicals, are of particular threat to humans in utero, when they can concentrate in the brain and have **neurotoxic** effects at critical stages of development. Animal tests reveal negative health effects on the liver, brain, and nervous system.

BFRs accumulate in the food chain but are also inhaled regularly, as they tend to collect on floors and electronics such as keyboards. BFRs can be passed from mothers to children through the placenta or breast milk.

Perfluorinates (PCFs)

PCFs are a group of chemicals that serve a variety of related purposes. They are oil and water resistant and nearly **impervious** to heat, which means they are active in a variety of nonstick, nonstain, and water-resistant products. They are some of the most prevalent chemicals in our

Did You Know?

Some studies have linked BPA to breast cancer, male reproductive system defects, miscarriage, immune system defects, and polycystic ovarian disease.

Did You Know?

In some populations, BFRs found in human breast milk have been doubling every five years.

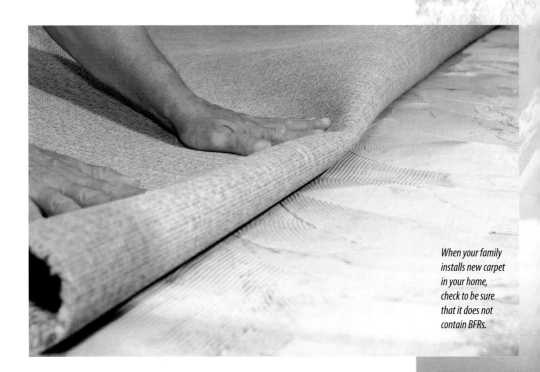

When your family installs new carpet in your home, check to be sure that it does not contain BFRs.

Safe Alternatives

Use stainless steel, cast iron, ceramic titanium, or porcelain-enameled cast iron cookware instead of PCF-treated cookware. If you do choose PCF cookware, be sure to keep the temperatures moderately low, and ventilate the room properly to protect those most sensitive—children and infants.

Avoid "wrinkle-free" and "stain-resistant" clothing if possible.

Avoid fast-food packaging, which can be a dual disaster for PCF exposure: often the packaging itself is treated with PCFs, and the high fat content of fast food risks absorption of the PCFs.

environment because of their persistence, and are found in humans and other living things all over the globe.

Studies show that certain kinds of fish break down less toxic fluorinated chemicals into PCFs that persist in the food chain and bioaccumulate until they reach humans.

The lining of microwave popcorn bags is a dangerous source of PCFs.

Concern has also been raised over PCF use in firefighting foams, which can save lives but leave hazardous chemicals to contaminate the sight of the fire.

The U.S. EPA has classified some PCFs as carcinogenic. They've also acknowledged that occupational exposure to some PCFs has been linked to an increase in bladder cancer. PCFs may cause birth defects, disrupt the immune system and proper thyroid function, and if exposure happens during pregnancy, they may cause developmental problems as well.

PCFs are found in waterproofed or stain-resistant products. These include outdoor clothing, wrinkle-resistant clothing, leather goods, floor waxes, and nonstick cookware. At extremely high temperatures, PCFs can break down into gases that, if inhaled or cooked into food and then ingested, can cause a rare disease called polymer fume fever. This risk is particular to cookware coated with PCFs. PCFs can also be found on the inside of food containers. The insides of popcorn bags are a particularly concentrated source of PCFs used to stop grease and butter from seeping through.

Formaldehyde

Formaldehyde is one chemical that probably almost everyone has heard of. What most people don't know, however,

> **Did You Know?**
>
> *Some manufacturers, such as 3M, have already agreed to phase out PCFs in their products explicitly because of evidence of the chemical's toxicity. Others have followed suit, but many PCF products remain. Eight U.S. manufacturers have agreed to reduce emissions of PCFs by 95 percent by 2010, and eliminate small amounts of PCFs in their products by 2015.*

Safe Alternatives

Avoid particleboard if at all possible. Use solid lumber or "exterior-grade" plywood—wood that hasn't been treated for special indoor purposes.

Stay away from paints, varnishes, and disinfectants with high concentrations of formaldehyde. Almost all these products have formaldehyde-free alternatives.

Don't allow smoking in closed environments. In fact, if possible avoid cigarette smoke at all costs, as it contains large amounts of formaldehyde (and other hazardous substances).

Real People

On July 19, 2007, Paul Stewart, a survivor of Hurricane Katrina who had received a camper from the U.S. Federal Emergency Management Agency (FEMA), presented the following words to the Government Reform and Oversight Committee of the U.S. House of Representatives:

On December 2, 2005 our FEMA camper was delivered. We had a friend, who had already returned to the area, meet FEMA and take possession of the camper and he immediately noticed that it had a very strong "new" smell. He told us that the fumes in the camper made his eyes burn so we instructed him to open all the windows and turn the heat on as my wife and I had already heard FEMA officials say that the campers needed to "air out" and had issued instructions on how to complete the process. Two days later we arrived home and I turned off the heat, but left the windows and exhaust vent open, the camper stayed that way for the next 4 months.

The first night we stayed in the camper my wife woke several times with difficulty breathing and a runny nose. She got up once and turned on the lights to discover that her runny nose was, in fact, a bloody nose.... We didn't know what was causing her bloody nose, or breathing issues and I was beginning to show symptoms of my own, which included, blurring eyes, scratchy throat, coughing, and runny nose. The symptoms continued for weeks and then months and finally we thought about just leaving, but at that point we were stuck because we ... just couldn't afford to move.

Then one morning I woke up to find our pet cockatiel lethargic, unresponsive, and unable to keep his balance....

We got to the veterinarian and he examined the bird telling us that we needed to get him out of the camper or it would kill him. He told us that the chemicals in the camper were too strong for the bird and his respiratory system could no take the toxins. The veterinarian told us that birds, much like children, breathe much more rapidly then adults and therefore take in not only more air, but more of the contaminants in the air. I asked the veterinarian what kind of contaminants could be inside a new camper. He told us that there are a lot of different chemicals that off gas from a newly constructed camper, but the birds' problem was probably being caused by formaldehyde. This was the first time it really hit us that our FEMA camper could actually be dangerous and thought if it could kill the bird, what was it doing to us?

... I then started researching what formaldehyde and the other contaminants inside my camper might be doing to our health and was shocked to discover that formaldehyde is listed as a carcinogen by the EPA.... I called FEMA and told them about the problems we were having and the first thing FEMA told us to do was "air out" the camper. I explained that I had already been "airing" out the camper for almost 2 months and it had done nothing to reduce the smell or our symptoms.... Our symptoms continued and I continued to tell the FEMA inspectors, who came out almost weekly, that the formaldehyde in our camper was making us sick.

industrial chemicals & health

Did You Know?

PCFs occur in highest amounts in children. This is likely the result of an increase in the PCFs used in clothing today, and the fact that children are more sensitive to them.

Did You Know?

Formaldehyde was used in insulating foam in the 1970s but banned in the 1980s in both the United States and Canada.

is that it's nearer to them than they think. Formaldehyde can be found indoors in places you might not expect: some carpet backing, curtains, flooring, and kitchen and bathroom cabinets contain formaldehyde. Even some bathroom products—certain lipsticks, toothpaste, and lotions—contain this chemical.

Formaldehyde is a volatile organic chemical (VOC). It has a tendency to off-gas into the environment, filling closed rooms with harmful toxic chemicals to be inhaled. Warm temperatures increase the rate of off-gassing. Formaldehyde is usually sensed immediately if it occurs in high enough concentrations. Red eyes and irritated throat, nose, and lungs are all symptoms of formaldehyde exposure. Some people even develop sensitivities over time because they are exposed to formaldehyde so often. These include skin and lung allergies, such as asthma.

Formaldehyde's largest source is standard MDF, or medium-density fiberboard. MDF is the most common particleboard used in building projects and can release formaldehyde into the air for over a year after it's installed.

Your toothpaste may contain formaldehyde! Check the ingredients of all lotions, cosmetics, shampoos, and toothpaste to make sure you are not exposing yourself to this chemical.

Common Toxic Chemicals and Their Health Effects

Formaldehyde fumes will accumulate in the indoor atmosphere of a building, causing "sick building syndrome." This includes a number of symptoms, such as irritation in the eyes, throat, and lungs. Mobile homes are especially at risk of formaldehyde exposure because there is little ventilation and smaller living space. Therefore, the U.S. Department of Housing and Urban Development requires mobile-home developers to use low-emitting materials and warn potential buyers of the risks of formaldehyde.

Why Don't the World's Governments Do Something?

If scientists know that industrial chemicals are dangerous, why are they continued to be use?

That's a good question. Partly, the people who run the world's governments don't want to face how dangerous these chemicals are because it would be expensive to find alternatives. Big businesses influence governments—and businesses are reluctant to spend money on finding new ways to do things when the old ways seem to be working. This may seem to be a silly way to think—after all, the owners of big businesses don't live in bubbles; they are exposed to dangerous chemicals just like the rest of us—but human beings are often reluctant to face unpleasant and uncomfortable truths. Nevertheless, the tide is start to turn; the people of the world are starting to take action.

Did You Know?

The U.S. EPA and the International Agency for Research in Cancer both classify formaldehyde as a probably human carcinogen.

Did You Know?

Aspartame, a chemical found in many diet drinks, tabletop sweeteners, and chewing gum, breaks down in the body into methanol, which further breaks down into formaldehyde, which over time can cause damage to the neurological and immune systems.

STRAIGHT FROM THE SOURCE

Another Perspective on Formaldehyde

The Formaldehyde Council's Web site offers this information on the chemical:

> Urinary tract infections afflict people worldwide. In the United States and Canada more than $1 billion is spent each year to treat them. The majority of the cases are treated using a derivative of formaldehyde (methenamine). While the chemical reactions may be a bit complex, the bottom line is that formaldehyde kills the infection. Antibiotics represent one alternative to using formaldehyde-based drugs, however, bacterial resistance develops using antibiotics. Bacteria are incapable of developing resistance to formaldehyde, so it remains the treatment of choice.
>
> Formaldehyde is used to create the enteric or hard capsules that are used to deliver drugs in the form of pills to millions of people worldwide every day. The formaldehyde-based pill coatings slow the dissolution of the capsule and promote maximum absorption of the medicine.
>
> Topical creams, cosmetics and personal hygiene products contain active ingredients that prevent the growth of potentially harmful bacteria. Some of these ingredients are derivatives of formaldehyde.
>
> Anyone suffering from coronary artery disease knows the horrible pain and anxiety that accompanies angina—the suffocating chest pain associated with lack of oxygen to the heart muscle. The nitroglycerin pills placed under the tongue that ease these attacks are made from a formaldehyde byproduct.

What Do You Think?

- What do you think is the motivation behind the Formaldehyde Council's presentation of this information?

- Do you think this summary is factually accurate? What could you do to find out?

- Do you think it presents a biased viewpoint? Why or why not?

Find Out More

To learn more about BFRs, PCFs, and formaldehyde, go to these Web sites:

www.noharm.org/us/bfr/issue

www.greenpeace.to/publications/Bayer%20report.pd

www.epa.gov/iaq/formalde.html

Here's what you need to know

- The current chemical regulation system is fundamentally flawed because it is reactive rather than precautionary.
- The chemical industry is a big business, and lobbyists are representatives for chemical companies that fight regulations that might diminish profits.
- Most countries try to keep consumers safe with proper regulations without losing economic growth that chemical companies offer. This is a difficult position to be in.
- In the past forty years, the United States has made significant progress toward better regulation of toxic chemicals, but it still has far to go before the burden of proof is placed upon chemical manufacturers.
- The Stockholm Convention prepared the way for the REACH program, the boldest move toward stricter regulation of chemicals to date. The REACH program, in particular, places Europe in a leadership position in the years to come. Time will tell how the governmental bodies of the EU will enforce these new laws.

Words to Understand

Reactive means you deal with a problem after it happens instead of preventing it.

A **precautionary** approach takes steps to prepare for possible dangers in the future.

If something is made into law then it is **enacted**.

Lobbyists are people who try to influence the decisions made by lawmakers.

An **ethical** decision takes into account ideas of right or wrong.

6
What Is the World Doing?

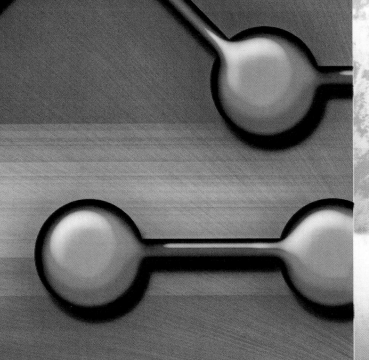

The world's attitude toward industrial chemicals is built on flawed reasoning. Currently, most governing bodies have decided to adopt an approach to toxic chemicals that is **reactive** rather than **precautionary**. A reactive approach waits for chemicals to cause harm and the public to raise the alarm before regulation is **enacted**.

This policy of "innocent until proven guilty" is effective in the industrialized world, where thousands of chemicals have been "grandfathered" into acceptable use without testing. What this means is that when chemical regulations began to emerge in the 1970s and 1980s, thousands of chemicals had already been in use for decades. Rather than testing every one for toxic qualities, the world's governments gave them the green light for continued use—placing the burden of responsibility upon regulatory and medical bodies to prove the chemical's harmfulness. Of course, in actuality this means that the real burden is placed upon the public. It is the uninformed consumer who is most at risk.

Big Business

What we are referring to when we use the term "big business" are large chemical corporations with billions of dollars invested in the interest of chemicals, not human beings. Don't be mistaken, these corporations are staffed by people like you and me, people with families who are just as much at risk from toxic chemicals as we are. The difference is that their jobs require them to act always in the interest of profit.

The representatives for big-business chemical companies who fight against regulation are called **lobbyists**. Lobbyists work carefully with government officials to keep regulation to a minimum, and their arguments carry a lot of weight because the chemical industry is such a vital part of the world economy. Every country must find a way to balance a plan for protecting the public with a plan to stimulate economic growth.

In addition to splitting the interests of government, the lobbyists for big-business chemicals have the advantage of

Timeline of Chemical Regulations

Here is a brief history of regulation from Ashton and Green's The Toxic Consumer:

1962—Rachel Carson publishes Silent Spring, the book that many environmentalists would argue launched the modern environmental movement.

1970—The U.S. Environmental Protection Agency (EPA) is established to help protect the nation's public health and environment through research, education, and enforcement of environmental laws, including the 1970 Clean Air Act.

1972—The EPA bans the use of DDT, a widely used pesticide found to be carcinogenic and accumulating in the food chain.

1974—Congress passes the Safe Drinking Water Act, setting health-based standards for the quality of the public water supply.

1976—Congress passes the Toxic Substances Control Act (TSCA), originally intended to assure the safety of new and existing chemicals, and to make information provided by chemical companies available to the public. Under the law, the EPA starts to review chemicals in 1979.

1977—The Consumer Products Safety Commission (CPSC) bans the fire retardant "tris" from use in children's sleepwear after it was shown to be a potent animal carcinogen.

1979—The EPA bans PCBs and two herbicides containing dioxins,.

1983—The Report of the UN Brundtland Commission illustrates the global concern for the state of the environment and popularizes the phrase "sustainable development." This is defined in the report as a way "to meet the needs of the present without compromising the ability of future generations to meet their own needs."

1988—The EPA bans the pesticide chlordane, used to treat over 30 million U.S. homes since the 1950s. Chlordane has been linked to childhood and adult cancers and a wide variety of respiratory and neurological disorders.

1989—The EPA bans commercial use of asbestos, but the decision is overturned by a federal appeals court in 1991. Since this time, no chemicals have been banned in the U.S.

1992—The UN convenes the Rio Earth Summit. The largest environmental conference held up to this date, the summit draws over 30,000 people, including more than 100 heads of state. The summit's "Declaration on Environment and Development" addresses the environmentally sound management of toxic chemicals, including the illegal international traffic in toxic and dangerous products. Additionally, the declaration states: "In order to protect the environment, the precautionary approach shall be widely applied by states according to their capabilities. Where there are threats of serious or irreversible damage, lack of full scientific certainty shall not be used as a reason for postponing cost-effective measures to prevent environmental degradation." This establishes the first legal precedent for the "Precautionary Principle," giving the environment the benefit of the doubt over unsustainable development.

2001—The UN adopts the Stockholm convention on Persistent Organic Pollutants (POPs). Signatories agree to phase out and limit production of 12 major POPs—toxic, bioaccumulative chemicals that can cause biological havoc. At present, over 140 countries have ratified the Convention. The U.S. has signed it, but has not yet ratified it.

2003—The European Commission proposes REACH, a new regulatory framework for the Registration, Evaluation and Authorization of Chemicals throughout Europe. Unlike U.S. law, REACH places the burden of proof on the manufacturer to demonstrate that chemicals are safe before they are placed on the market. REACH was formally adopted by the EU in 2006 and went into force in mid-2007.

Did You Know?

In Europe, especially Germany, chemical industries are among the largest companies. Together they generate about 3.2 million jobs in more than 60,000 companies. In 2005, the top-ten largest companies in Europe generated more than $250 billion (U.S. dollars) in sales.

Did You Know?

The chemical industry is been concentrated in three areas of the world: Western Europe, North America, and Japan. The European Union is the largest producer, followed by the United States and Japan.

working in a reactionary system, whereby extensive evidence must be produced before a chemical will be banned or restricted. Because there is a system of self-regulation in countries such as the United States, many companies face the **ethical** problem of properly conducting tests that could very well ban their product from the market. There is no small amount of money at stake when it comes to the chemical industry. To put things in perspective, the top ten chemical companies amassed just over 320 billion dollars in the year 2005 alone.

The United States and Chemical Regulation

The United States has had difficulty passing laws that firmly reverse the burden of proof and place it upon industry instead of government and the public. Unfortunately, this is still the case today. The situation becomes increasingly complicated when one considers the number of chemicals in production—over 10,000 of which are in consumer products used every day—and the potentially harmful interactions these chemicals might have in our bodies. In addition, most cancers take twenty or more years to develop, further distancing chemical culprits from their victims. To the United States' credit, they have established three organizations that attempt to regulate toxic chemicals in different sectors of society: the EPA, the FDA, and the CPSC.

Environmental Protection Agency (EPA)

Since its inception in 1970, the EPA has had the enormous task of acting as the main regulating body for environmental concerns in one of the fastest-growing countries in the world. It is in fact speed that seems to be a recurring theme in chemical regulation discussions: the regulatory agencies (such as the EPA) simply cannot keep up with the production of new chemicals every year. In 1976, the EPA was successful in passing the Toxic Substances Control Act. However, they were unable to thoroughly test the 60,000 chemicals already in use, and so only about 200 were tested and the rest were accepted without question. What is frightening is that these 60,000 relatively untested chemicals account for 99 percent of the chemicals in use today.

The EPA is aware of its insufficiency in obtaining toxicity reports, and in 1998 launched a voluntary program called the High Production Volume (HPV) Challenge. The program aimed to provide basic toxicity data on the nearly 28,000 chemicals produced at rates over one million pounds per year, and then make this data available to the public. It has been somewhat successful, but suffers because of its voluntary nature—meaning chemical companies are not required to participate.

The Food and Drug Administration (FDA)

The FDA monitors cosmetics and drugs sold in the United States, which means they interact with the chemical industry regularly. However, as is usually the case with the chemical industry, the FDA has limited say over how much testing is required before a product can be put on the market. When it comes to drugs, the FDA has firmly established a rule that all drugs must have adequate test data before they can be sold. This means that a chemical must be tested by the company selling it—putting the burden of proof, for once, on the chemical industry.

Cosmetics are a different story. They are "self-regulated" by the chemical industry—it is the manufacturer's job to insure the product is safe, but there are no requirements on how this must be accomplished.

Also frightening is that the FDA does not require "indirect additives" to be included on product labels. Indirect additives are chemicals added to packaging or containers of products. The vast majority of toxic chemicals that leach into food and cosmetics are placed in this "indirect additive" category, which means they go completely unnoticed.

The Consumer Products Safety Commission (CPSC)

The CPSC was created in 1972 to protect the public from hazards that arise in specific consumer products. The CPSC has a history of addressing one or two products at a time that stand out as threats to public health. They've banned everything from faulty microwaves to baby teethers. Chemical-related bans, however, have been very rare in the CPSC history.

Even so, there have been successful instances of CPSC targeting chemical toxins in the market. One instance was their 1977 ban of "tris," a fire retardant used in children's sleepwear that was found to be a carcinogen in animals.

A second example was 1998's controversy over children's chew toys that contained a type of phthalate. Parents petitioned the CPSC to ban the product line and in response CPSC did not impose a ban, but did request that the industry remove phthalates from chew toys until more scientific work was done. In contrast, both Japan and the European Union have established strict bans of phthalates in children's toys.

The Stockholm Convention

This convention was a major achievement of global chemical regulation. It followed in the footsteps of the Rio Earth Summit of 1992, and took steps to put into action many of the environmental-protection principles talked about for some time. The convention became known as the Persistent Organic Pollutants (or POP) convention because it focused on identifying toxic chemicals that were accumulating in the environment and in our bodies. The convention identified twelve of the worst POPs and called them "the dirty dozen." Over 140 countries signed the agreement to ban these chemicals, including many of the world powers.

The Stockholm Convention was a big step forward, but it won't be enough to keep toxic chemicals out of our food and our environment. The convention only addresses twelve compounds, after all—many of which had already been banned in the 1970s by developed countries such as the United States, Canada, Great Britain, and other European nations.

Registration, Evaluation, and Authorization of Chemicals (REACH)

REACH came about as a result of continued concern over toxic chemicals being traded between European Union (EU) countries. In 1998, the EU convened and proposed that new legislation was needed to regulate chemicals. The goal at that time was to study and test the thousands of chemicals already on the market, and to reverse the burden of proof so that chemical manufacturers were responsible for proving a chemical safe—instead of the public having to prove it harmful.

This led to the "Chemicals White Paper," which in turn led to a new vision: REACH. This new vision would put

Because we live in a global community that is increasingly shaped by both an international economy and environmental issues that respect no boundaries, the people and leaders of the world must work together in order to research, regulate, and restrict the use of industrial chemicals.

into practice the goals set forth by the "Chemicals White Paper," and would do so on these principles:

- To phase out the use of chemicals of high concern—persistent, bioaccumulative, toxic chemicals (PBTs), and endocrine-disrupting chemicals (EDCs)
- To find safer alternatives for high-risk chemicals
- The Right to Know (RTK): the public has a right to know which chemicals are in the products they use
- "No data, no market": if a manufacturer does not have information proving a chemical safe, it is not ready to be sold in the market

Of all the global legislation concerning toxic chemicals, REACH, which went into effect on June 1, 2007, seems to be at the forefront. Time will tell how effective the program will be in reducing toxic chemicals in the world. Already, U.S. and European representatives from chemical companies have raised complaints that REACH is an unfair obstacle to global trade.

What the Future Looks Like

We can reasonably expect steps to be taken toward better labeling of chemically treated products, as well as the eventual phasing out of PBT chemicals. What is less likely, however, is that a chemical industry that has gone unregulated for so many years will immediately begin regulating itself. It will take time and effort to establish stricter codes for chemical manufacturers to follow.

STRAIGHT FROM THE SOURCE

REACH in Brief

According to the European Commission's REACH Web site:

[REACH's] two most important aims are to improve protection of human health and the environment from the risks of chemicals while enhancing the competitiveness of the EU chemicals industry.

REACH is based on the idea that industry itself is best placed to ensure that the chemicals it manufactures and puts on the market in the EU do not adversely affect human health or the environment. This requires that industry has certain knowledge of the properties of its substances and manages potential risks. Authorities should focus their resources on ensuring industry are meeting their obligations and taking action on substances of very high concern or where there is a need for community action.

What Do You Think?

- According to this document, what are REACH's two most important aims?
- Do you think both aims can be achieved at the same time?
- Do you think REACH's goals are big enough? Why or why not?

Find Out More

To find out more about what the world is doing about industrial chemicals, check out these Web sites:

www.chemicalspolicy.org

ec.europa.eu/environment/chemicals/reach/reach_intro.htm

www.epa.gov/tri/

For More Information on Health & the Environment

Books

Gordon, Bruce, Richard Mackay and Eva Rehfuess. *Inheriting the World: The Atlas of Children's Health and the Environment.* World Health Organization, 2004.

Ho, Mun S. and Chris P. Nielsen, eds. *Clearing the Air: the Health and Economic Damages of Air Pollution in China.* Cambridge, MA: MIT Press, 2007.

Kusinitz, Marc. *Poisons and Toxins.* New York: Chelsea House Publications, 1992.

MacDonald, John J. *Environments for Health.* New York: Earthscan Publications, 2006.

McCally, Michael, ed. *Life Support: The Environment and Human Health.* Cambridge, MA: MIT Press, 2002.

Nadakavukaren, Anne. *Our Global Environment: a Health Perspective.* 6th edition. New York: Waveland Press, 2005.

Nakaya, Andrea C. *Is Air Pollution a Serious Threat to Health?* New York: Greenhaven Press, 2004.

Netzley, Patricia D. *Contemporary Issues: Issues in the Environment.* New York: Lucent Books, 1997.

Vesley, Donald. Human *Health and the Environment: A Turn of the Century Perspective.* New York: Springer, 1999.

Web Sites

Air Pollution
health.nih.gov/result.asp/19

For More Information on Health & the Environment

Air Quality
www.epa.gov/airnow/

CDC: Environmental Health
www.cdc.gov/Environmental/

EPA: Environmental Kids Club
www.epa.gov/kids/

The Green Squad
www.nrdc.org/greensquad/intro/intro_1.asp

Health and Environmental Linkages Initiative
www.who.int/heli/en/

International Year of Sanitation
www.who.int/water_sanitation_health/hygiene/iys/about/en/index3.html

Kids for Saving Earth
www.kidsforsavingearth.org/index_low.html

Library of Congress Environmental; Photographs
memory.loc.gov/ammem/award97/icuhtml/aephome.html

Public Health and the Environment
www.who.int/phe/en/

Teen Ink
www.teenink.com/Environment/index.php

Toxic Household Cleaners
www.tutorials.com/08/0858/0858.asp

United Nations Population Fund
www.unfpa.org/

Water and Sanitation Quiz
www.unicef.org/voy/explore/wes/1883_wes_quiz.php

Glossary of Environmental Health–Related Terms

When you're reading about environmental health, especially in some of the more technical government reports, you may encounter many unfamiliar medical terms. This glossary can help you better understand the words scientists and other experts use when talking about the effects of environmental pollution on human health.

Absorption
The process of taking in; for a person or an animal, this refers to a substance getting into the body through the eyes, skin, stomach, intestines, or lungs. Chemicals can be absorbed into the bloodstream after breathing or swallowing. Chemicals can also be absorbed through the skin, into the bloodstream, and then transported to other organs. Not all chemicals breathed, swallowed, or touched are absorbed.

Acute
Occurring over a short time, usually a few minutes or hours. An acute exposure only lasts for up to 14 days; it can result in short-term or long-term health effects. An acute effect happens a short time after exposure.

Additive Effect
The body's response to exposure to multiple substances that equals the sum of responses of all the individual substances added together.

Adverse Health Effect
A change in body function or cell structure that might lead to disease or health problems.

Ambient
Surrounding. Ambient air usually means outdoor air (as opposed to indoor air).

Glossary of Environmental Health–Related Terms

Analyte
A chemical for which a sample (such as water, air, blood, urine, or another substance) is tested and measured in the laboratory. For example, if the analyte is mercury, the laboratory test will determine the amount of mercury in the sample.

Antagonistic Effect
A biologic response to exposure to multiple substances that is less than would be expected if the known effects of the individual substances were added together.

Aquatic Ecosystem
A community of organisms that live together in a body of water and are interdependent.

Aquifer
A geological formation where the spaces between rock particles, sand, or gravel are completely filled with water. Water pumped from aquifers is referred to as "groundwater".

Background Level
A typical or average level of a chemical in the environment. Background often refers to naturally occurring or uncontaminated levels. Background levels in one region of the world may be different than those in other areas.

Bedrock
The solid rock underneath surface soils.

Biodegradation
Decomposition or breakdown of a substance through the action of microorganisms (such as bacteria or fungi) or other natural, physical processes (such as sunlight).

Biologic Indicators of Exposure Study
A study that uses medical tests and other markers of exposure in human body fluids or tissues to confirm human exposure to a hazardous substance.

Biological Monitoring
Measuring chemicals, hormone levels, or other substances in biological materials (blood, urine, breath, or hair) as a measure of chemical exposure and health in humans or animals. A blood test for lead is an example of biological monitoring.

Biologic Uptake
The transfer of substances from the environment to plants, animals, and humans.

Biomedical Testing
Testing of persons to find out whether a change in a body function might have occurred because of exposure to a hazardous substance in the environment.

Biota
Plants and animals in an environment. Some of these plants and animals might be sources of food, clothing, or medicines for people.

Body Burden
The total amount of a chemical in the body. Some chemicals build up in the body because they are stored in body organs like fat or bone or are eliminated very slowly.

Cancer
Any one of a group of diseases that occur when cells in the body become abnormal and grow or multiply out of control.

Cancer Risk
The theoretical risk for getting cancer if exposed to a substance every day for 70 years (a lifetime exposure). The true risk might be lower.

Carcinogen
A substance that causes cancer.

Case Study
A medical evaluation of one person or a small group of people to gather information about specific health conditions and past exposures.

Case-Control Study
A study in which a group of people with a disease (cases) are compared to people without the disease (controls) to see if their past exposures to chemicals or other risk factors were different.

Central Nervous System (CNS)
The part of the nervous system that includes the brain and the spinal cord.

Chronic
Occurring over a long period of time, several weeks, months, or years.

Chronic Exposure
Contact with a substance that occurs over a long time (more than a year).

Cluster Investigation
A review of an unusual number, real or perceived, of health events (for example, reports of cancer) grouped together in time and location. Cluster investigations are designed to confirm case reports; determine whether they represent an unusual disease occurrence; and, if possible, explore possible causes and contributing environmental factors.

Cohort Study
A study in which a group of people with a past exposure to chemicals or other risk factors are followed over time and their disease experience compared to that of a group of people without the exposure.

Comparison Value (CV)
Calculated concentration of a substance in air, water, food, or soil that is unlikely to cause harmful (adverse) health effects in exposed people. The CV is used as a screening level during the public health assessment process. Substances found in amounts greater than their CVs might be selected for further evaluation in the public health assessment process.

Composite Sample
A sample which is made by combining samples from two or more locations. The sample can be of water, soil, or another substance found in the environment.

Concentration
The amount of one substance dissolved or contained in a given amount of another substance. For example, sea water has a higher concentration of salt than fresh water does.

Contaminant
Any substance found somewhere (for example, the environment, the human body, or food) where it is not normally found. Contaminants are usually referred to in a negative sense and include substances that spoil food, pollute the environment, or cause other adverse effects.

Delayed Health Effect
A disease or an injury that happens as a result of exposure that might have occurred in the past.

Dermal
Having to do with the skin. For example, dermal absorption means absorption through the skin.

Dermal Contact
Touching the skin.

Detection Limit
The smallest amount of substance that a laboratory test can reliably measure in a sample of air, water, soil, or other medium.

Dose
The amount of substance to which a person is exposed. Dose is a measurement of exposure and is often expressed as milligram (amount) per kilogram (a measure of body weight) per day (a measure of time) when people eat or drink contaminated water, food, or soil. In general, the greater the dose, the greater the likelihood of an effect. An "exposure dose" is how much of a substance is encountered in the environment. An "absorbed dose" is the amount of a substance that actually got into the body through the eyes, skin, stomach, intestines, or lungs. For radioactive chemicals, dose is the amount of energy from radiation that is actually absorbed by the body. This is not the same as the measurement of the amount of radiation in the environment.

Dose-Response Relationship
The relationship between the amount of exposure to a substance and the resulting changes in body function or health.

Environmental Media and Transport Mechanism
Environmental media include water, air, soil, plants, and animals. Transport mechanisms move contaminants from the source to points where human exposure can occur. The environmental media and transport mechanism is the second part of an exposure pathway.

EPA
United States Environmental Protection Agency.

Epidemiology
The study of the occurrence and causes of health effects in human populations. An epidemiological study often

compares two groups of people who are alike except for one factor such as exposure to a chemical or the presence of a health effect. The investigators try to determine if the factor is associated with the health effect.

Exposure
Contact with a chemical by swallowing, breathing, or direct contact (such as through the skin or eyes). Exposure may be either short term (acute) or long term (chronic).

Exposure Assessment
The process of finding out how people come into contact with a hazardous substance, how often and for how long they were in contact with the substance, and how much of the substance they were in contact with.

Exposure-Dose Reconstruction
A method of estimating the amount of people's past exposure to hazardous substances. Computer and approximation methods are used when past information is limited, not available, or missing.

Exposure Investigation
The collection and analysis of information from an environmental site and biologic tests to determine whether people have been exposed to hazardous substances.

Exposure Pathway
The route a substance takes from its source (where it began) to its end point (where it ends), and how people can come into contact with (or get exposed to) it along the way. An exposure pathway has five parts: a source of contamination (such as an abandoned business); an environmental media and transport mechanism (such as movement through groundwater); a point of exposure (such as a private well); a route of exposure (eating, drinking, breathing, or touching), and a receptor population (people potentially or actually exposed). When all five parts are present, the exposure pathway is termed a completed exposure pathway.

Exposure Registry
A system of the ongoing follow-up of people who have had documented environmental exposures.

Feasibility Study (FS)
A study that compares different ways to clean up a contaminated site. The feasibility study recommends one or more actions to remediate the site.

Geographic Information System (GIS)
A mapping system that uses computers to collect, store, manipulate, analyze, and display data. For example, GIS can show the concentration of a contaminant within a community in relation to points of reference such as streets and homes.

Gradient
The change in a property over a certain distance. For example, lead can accumulate in surface soil near a road due to automobile exhaust. As you move away from the road, the amount of lead in the surface soil decreases. This change in the lead concentration with distance from the road is called a gradient.

Groundwater
Water beneath the earth's surface in aquifers (as opposed to surface water)

Half-Life
The time it takes for half the original amount of a substance to disappear. In the environment, the half-life is the time it takes for half the original amount of a substance to disappear when it is changed to another chemical by bacteria, fungi, sunlight, or other chemical processes. In the human body, the half-life is the time it takes for half the original amount of the substance to disappear, either by being changed to another substance or by leaving the body. In the case of radioactive material, the half life is the amount of time necessary for one half the initial number of radioactive atoms to change or transform into another

atom (that is normally not radioactive). After two half lives, 25% of the original number of radioactive atoms remain.

Hazard
A source of potential harm from past, current, or future exposures.

Hazardous Waste
Potentially harmful substances that have been released or discarded into the environment.

Health Assessment for Contaminated Sites
Determination of actual or possible health effects due to environmental contamination or exposure. It includes a health-based interpretation of all the information known about the situation. The information may come from site investigations (environmental sampling and studies), exposure assessments, risk assessments, biological monitoring, or health effects studies. The health assessment is used to advise people on how to prevent or reduce their exposures, to determine what action to take to improve the situation, or the need for additional studies.

Health Effects Studies (related to contaminants)
Studies of the health of people who may have been exposed to contaminants. They include, but are not limited to, epidemiological studies, reviews of the health status of people in exposure or disease registries, and doing medical tests.

Health Registry
A record of people exposed to a specific substance (such as a heavy metal), or having a specific health condition (such as cancer or a communicable disease).

Incidence
The number of new cases of disease in a defined population over a specific time period

Ingestion
Swallowing (such as through eating or drinking). After ingestion, chemicals may be absorbed into the blood and distributed throughout the body.

Inhalation
Breathing. People can take in chemicals by breathing contaminated air.

Interim Remedial Measure (IRM)
An action taken at a contaminated site to reduce the chances of human or environmental exposure to site contaminants. Interim remedial measures are planned and carried out before comprehensive remedial studies. They can prevent additional damage during the study phase, but don't interfere in any way with the need to develop a complete remedial program. An example of an interim remedial measure is removing drums of chemicals to a storage facility from a site that has drums sitting in an empty field.

In Vitro
In an artificial environment outside a living organism or body. For example, some tests are done on cell cultures or slices of tissue grown in the laboratory, rather than on a living animal.

In Vivo
Within a living organism or body. For example, when scientific research is done on whole animals, such as rats or mice

Latency period
The period of time between exposure to something that causes a disease and the onset of the health effect. Cancer caused by chemical exposure may have a latency period of 5 to 40 years.

Leaching
As water moves through soils or landfills, chemicals in the soil may dissolve in the water, thereby contaminating the groundwater. This is called leaching.

Maximum Contaminant Level (MCL)
The highest (maximum) level of a contaminant allowed to go uncorrected by a public water system under federal or state regulations. Depending on the contaminant, allowable levels might be calculated as an average over time or might be based on individual test results. Corrective steps are implemented if the MCL is exceeded.

Media
Elements of a surrounding environment that can be sampled for contamination: usually soil, water, or air. Plants as well as humans (when sampling body substances such as blood or urine) and animals (such as sampling fish to update fish consumption advisories) can also be considered media. The singular of "media" is "medium."

Metabolism
All the chemical reactions that enable the body to work. For example, food is metabolized (chemically changed) to supply the body with energy. Chemicals can be metabolized by the body and made either more or less harmful.

Metabolite
Any product of metabolism.

Morbidity
Illness or disease. A morbidity rate for a certain illness is the number of people with that illness divided by the number of people in the population from which the illnesses were counted.

Mortality
Death. Usually the cause (a specific disease, a condition, or an injury) is stated along with this term.

Glossary of Environmental Health–Related Terms

Mutagen
A substance that causes mutations (genetic damage).

Mutation
A change (damage) to the DNA, genes, or chromosomes of living organisms.

Odor Threshold
The lowest concentration of a chemical that can be smelled. Different chemicals have different odor thresholds. Also, some people can smell a chemical at lower concentrations than others can.

Organic
Generally considered as originating from plants or animals, and made primarily of carbon and hydrogen. Scientists use the term organic to mean those chemical compounds which are based on carbon.

Permeability
The property of permitting liquids or gases to pass through. A highly permeable soil, such as sand, allows a liquid to pass through quickly. Clay has a low permeability.

Persistence
The quality of remaining for a long period of time (such as in the environment or the body). Persistent chemicals (such as DDT and PCBs) are not easily broken down.

Plume
An area of chemicals moving away from its source in a long band or column. A plume, for example, can be a column of smoke from a chimney or chemicals moving with groundwater.

Point of Exposure
The place where someone can come into contact with a substance present in the environment (see **Exposure Pathway**).

Population
A group or number of people living within a specified area or sharing similar characteristics (such as occupation or age).

Prevalence
The number of existing disease cases in a defined population during a specific time period.

Prevention
Actions that reduce exposure or other risks, keep people from getting sick, or keep disease from getting worse.

Protocol
The detailed plan for conducting a scientific procedure. A protocol for measuring a chemical in soil, water, or air describes the way in which samples should be collected and analyzed.

Radioisotope
An unstable or radioactive isotope of an element that can change into another element by giving off radiation.

Radionuclide
Any radioactive isotope of any element.

Receptor population
People who could come into contact with hazardous substance.

Registry
A systematic collection of information on persons exposed to a specific substance or having specific diseases.

Remedial Investigation (RI)
An in-depth study (including sampling of air, soil, water, and waste) of a contaminated site needing remediation to determine the nature and extent of contamination. The remedial investigation (RI) is usually combined with a feasibility study (FS).

Glossary of Environmental Health-Related Terms

Remediation
Correction or improvement of a problem, such as work that is done to clean up or stop the release of chemicals from a contaminated site. After investigation of a site, remedial work may include removing soil and/or drums, capping the site, or collecting and treating the contaminated fluids.

Risk
Risk is the possibility of injury, disease, or death. For example, for a person who has measles, the risk of death is one in one million.

Risk Assessment
A process which estimates the likelihood that exposed people may have health effects. The four steps of a risk assessment are: hazard identification (Can this substance damage health?); dose-response assessment (What dose causes what effect?); exposure assessment (How and how much do people come into contact with it?); and risk characterization (combining the other three steps to characterize risk and describe the limitations and uncertainties).

Risk Management (or Reduction)
The process of deciding how and to what extent to reduce or eliminate risk factors by considering the risk assessment, engineering factors (Can procedures or equipment do the job? For how long and how well?), social, economic, and political concerns.

Route of Exposure
The way in which a person may contact a chemical substance. For example, drinking (ingestion) and bathing (skin contact) are two different routes of exposure to contaminants that may be found in water. See **Exposure**.

Safe
Free from harm or risk. Exposure to a chemical usually has some risk associated with it, although the risk may

be very small. However, many people use the word safe to mean something that has a very low risk or one that is acceptable to them.

Sample
A portion or piece of a whole. A selected subset of a population or subset of whatever is being studied. For example, in a study of people the sample is a number of people chosen from a larger population. (See **Population**.) An environmental sample (for example, a small amount of soil or water) might be collected to measure contamination in the environment at a specific location.

Sample Size
The number of units chosen from a population or an environment.

Solubility
The largest amount of a substance that can be dissolved in a given amount of a liquid, usually water. For a highly water-soluble compound, such as table salt, a lot can dissolve in water. Motor oil is only slightly soluble in water.

Solvent
A liquid capable of dissolving or dispersing another substance (for example, acetone or mineral spirits).

Source of Contamination
The place where a hazardous substance comes from, such as a landfill, waste pond, incinerator, storage tank, or drum. A source of contamination is the first part of an **Exposure Pathway**.

Special Populations
People who might be more sensitive or susceptible to exposure to hazardous substances because of factors such as age, occupation, sex, or behaviors (for example, cigarette smoking). Children, pregnant women, and older people are often considered special populations.

Stakeholder
A person, group, or community who has an interest in activities at a hazardous waste site.

Statistics
A branch of mathematics that deals with collecting, reviewing, summarizing, and interpreting data or information. Statistics are used to determine whether differences between study groups are meaningful.

Superfund
The United States' federal and state program that investigates and cleans up inactive, hazardous waste sites.

Surface Water
Water on the surface of the earth, such as in lakes, rivers, streams, ponds, and springs.

Synergistic effect
A biologic response to multiple substances where one substance worsens the effect of another substance. The combined effect of the substances acting together is greater than the sum of the effects of the substances acting by themselves.

Target Organ
An organ (such as the liver or kidney) that is specifically affected by a toxic chemical.

Teratogen
A substance that causes defects in development between conception and birth.

Toxic Agent
Chemical or physical (for example, radiation, heat, cold, microwaves) agents that, under certain circumstances of exposure, can cause harmful effects to living organisms.

Toxicology

The study of the harmful effects of substances on humans or animals.

Tumor

An abnormal mass of tissue that results from excessive cell division that is uncontrolled and progressive. Tumors perform no useful body function. Tumors can be either benign (not cancer) or malignant (cancer).

Volatile

Evaporating readily at normal temperatures and pressures. The air concentration of a highly volatile chemical can increase quickly in a closed room.

Volatile Organic Compound (VOC)

An organic chemical that evaporates readily. Petroleum products such as kerosene, gasoline, and mineral spirits contain VOCs. Chlorinated solvents, such as those used by dry cleaners or contained in paint strippers, are also VOCs.

Bibliography

Ashton, Karen and Elizabeth Salter Green. *The Toxic Consumer: Living Healthy in a Hazardous World.* New York: Sterling, 2008.

Baird, Colin and Michael Cann. *Environmental Chemistry.* 3rd edition. New York: W.H. Freeman and Company, 2005.

Blanc, Paul D. *How Everyday Products Make People Sick: Toxins at Home and in the Workplace.* Berkeley: University of California Press, 2007.

The Bopal Medical Appeal. "What Happened in Bopal?" www.bhopal.org/whathappened.html.

Brown, Phil. *Toxic Exposures: Contested Illnesses and the Environmental Health Movement*: New York: Columbia University Press, 2007.

European Commission. *REACH: In Brief.* 2006. ec.europa.eu/environment/chemicals/reach/pdf/2007_02_reach_in_brief.pdf

Formaldehyde Council. "Health & Safety: Everyday Benefits." www.formaldehyde.org/healthSafety/benefits.html

Kerns, Thomas. *Environmentally Induced Illnesses: Ethics, Risk Assessment and Human Rights.* Jefferson, NC: McFarland & Company, 2001.

Scott-Clark, Cathy and Adrian Levy. "Spectre Orange." *The Guardian*, March 29, 2003.

Steinman, David and R. Michael Wisner. *Living Healthy in a Toxic World.* New York: Perigree, 1996.

Stewart, Paul. "Testimony Before the Government Reform and Oversight Committee, U.S. House of Representatives." July 19, 2007. oversight.house.gov/documents/20070719103124.pdf

Index

acetone 22-23
Agent Orange 52, 56
alchemy 11-12

bioaccumulation 8, 17-18, 25, 38, 43, 51, 69, 72, 83, 88
BFRs 20, 29, 66, 69-71, 79
Bisphenol A (BPA) 66, 68-71
body burden 26, 28-29, 33, 35, 61, 94
breast cancer 19, 52, 58, 71

chloracne 52, 55, 57

Environmental Protection Agency (EPA) 8, 11, 13, 23-25, 45, 47, 55, 59, 62, 73, 75, 77, 83-85, 97
epidemiology 38, 39, 46, 55, 97

DDT 18, 29, 38, 43-45, 83
dioxins 18, 52-58, 63-65, 83

Food and Drug Association (FDA) 84-85
formaldehyde 66, 73, 75-79

hazardous waste 8, 9, 14, 24-25, 58, 61, 64, 100

Industrial Revolution 8, 55
insecticide 38, 40-42, 44-45, 47, 62

landfill 22-23
leachate 9, 20, 22
lipophilic 8, 17, 32-33

organic chemicals 9, 11, 16-18, 20, 38, 40, 42, 45, 59, 76, 83, 86
organophospates 38, 45-46

Parkinson's Disease 38, 48
PCBs 18, 29, 52-54, 58-59, 61, 63-64, 83
pesticide 34-35, 38-51, 54, 63, 83
Persistant Organic Pollutant (POP) 18, 83
placenta 26, 33, 69, 71
polymer 9, 13, 73
pthalates 34, 52-54, 61-63, 86

Silent Spring 45, 83
Stockholm Convention 18, 70, 80, 83, 86

toxicology 38, 39, 46-47, 62

Volatile Organic Compound (VOC) 20-21, 76

Picture Credits

Dan Shea/Vietnam Agent Orange Relief and Responsibility Campaign
  p. 57
Dreamstime
  Amgreen: p. 22
  Cherick: p. 41
  Eraxion: p. 28
  Farek: p. 69
  FullValue: p. 71
  Hallgerd: p. 72
  Koele: pp. 8–9
  Koko77: p. 15
  Ljupco: pp. 66–67
  MadArtists: p. 30
  Meg380: p. 26–27
  Milosluz: p. 76
  Newphotoservice: pp. 52–53
  Oguzaral: pp. 21, 34
  Petarneychev: p. 17
  Rcmathiraj: p. 54
  Soupstock: pp. 38–39
  Webking: p. 10
  Ximagination: p. 33
  Y-ntousiopoulus: p. 16

Jupiter Images
  pp. 80–81, 87
USDA
  p. 44

To the best knowledge of the publisher, all other images are in the public domain. If any image has been inadvertently uncredited, please notify Harding House Publishing Service, Vestal, New York 13850, so that rectification can be made for future printings.

About the Author

Zachary Chastain is a writer currently living in Binghamton, New York, where he continues to do journalistic work on issues of environment and nutrition.

About the Consultant

Elise DeVore Berlan, MD, MPH, FAAP, is a faculty member of the Division of Adolescent Health at Nationwide Children's Hospital and an Assistant Professor of Clinical Pediatrics at The Ohio State University College of Medicine. She completed her Fellowship in Adolescent Medicine at Children's Hospital Boston and obtained a Master's Degree in Public Health at the Harvard School of Public Health. Dr. Berlan completed her residency in pediatrics at the Children's Hospital of Philadelphia, where she also served an additional year as Chief Resident. She received her medical degree from the University of Iowa College of Medicine. Dr. Berlan is board certified in Pediatrics and board eligible in Adolescent Medicine. She provides primary care and consultative services in the area of Young Women's Health, including gynecological problems, concerns about puberty, reproductive health services, and reproductive endocrine disorders.

Anne Nadakavukaren is the author *Our Global Environment: A Health Perspective*. She has taught at Illinois State University, and she also served on Illinois's Structural Pest Control Advisory Council, an advisory group to the Department of Public Health. She is currently a member of the Illinois Low-Level Radioactive Waste Advisory Taskforce.

LESTER B. PEARSON HIGH SCHOOL
RESOURCE CENTRE
1433 HEADON ROAD
BURLINGTON, ONTARIO
L7M 1V7

DATE DUE

1 3 OCT 2011

JAN 1 3 2017

Return Material Promptly